Moscow Lectures

Volume 3

More information about this series at http://www.springer.com/series/15875

Sergey M. Natanzon

Complex Analysis, Riemann Surfaces and Integrable Systems

NATIONAL RESEARCH
UNIVERSITY

Skoltech
Skolkovo Institute of Science and Technology

Springer

Sergey M. Natanzon
HSE University
Moscow, Russia

Translated from the Russian by Natalia Tsilevich.: Originally published as Комплексный анализ, римановы поверхности и интегрируемые системы by МССМЕ, 2018.

ISSN 2522-0314   ISSN 2522-0322 (electronic)
Moscow Lectures
ISBN 978-3-030-34642-3 ISBN 978-3-030-34640-9 (eBook)
https://doi.org/10.1007/978-3-030-34640-9

Mathematics Subject Classification (2010): 30C35, 30F10, 32G15, 37K10, 37K20

Cover illustration: https://www.istockphoto.com/de/foto/panorama-der-stadt-moskau-gm490080014-75024685, with kind permission

This Springer imprint is published by the registered company Springer Nature Switzerland AG.
The registered company address is: Gewerbestrasse 11, 6330 Cham, Switzerland

Preface to the Book Series *Moscow Lectures*

You hold a volume in a textbook series of Springer Nature dedicated to the Moscow mathematical tradition. Moscow mathematics has very strong and distinctive features. There are several reasons for this, all of which go back to good and bad aspects of Soviet organization of science. In the twentieth century, there was a veritable galaxy of great mathematicians in Russia, while it so happened that there were only few mathematical centers in which these experts clustered. A major one of these, and perhaps the most influential, was Moscow.

There are three major reasons for the spectacular success of Soviet mathematics:

1. Significant support from the government and the high prestige of science as a profession. Both factors were related to the process of rapid industrialization in the USSR.
2. Doing research in mathematics or physics was one of very few intellectual activities that had no mandatory ideological content. Many would-be computer scientists, historians, philosophers, or economists (and even artists or musicians) became mathematicians or physicists.
3. The Iron Curtain prevented international mobility.

These are specific factors that shaped the structure of Soviet science. Certainly, factors (2) and (3) are more on the negative side and cannot really be called favorable but they essentially came together in combination with the totalitarian system. Nowadays, it would be impossible to find a scientist who would want all of the three factors to be back in their totality. On the other hand, these factors left some positive and long lasting results.

An unprecedented concentration of many bright scientists in few places led eventually to the development of a unique "Soviet school". Of course, mathematical schools in a similar sense were formed in other countries too. An example is the French mathematical school, which has consistently produced first-rate results over a long period of time and where an extensive degree of collaboration takes place. On the other hand, the British mathematical community gave rise to many prominent successes but failed to form a "school" due to a lack of collaborations. Indeed, a school as such is not only a large group of closely collaborating individuals but also a group knit tightly together through student-advisor relationships. In the USA, which is currently the world leader in terms of the level and volume of mathematical research, the level of mobility is very high, and for this reason there are no US mathematical schools in the Soviet or French sense of the term. One can talk not only about the Soviet school of mathematics but also, more specifically, of the Moscow, Leningrad, Kiev, Novosibirsk, Kharkov, and other schools. In all these places, there were constellations of distinguished scientists with large numbers of students, conducting regular seminars. These distinguished scientists were often not merely advisors and leaders, but often they effectively became spiritual leaders in a very general sense.

A characteristic feature of the Moscow mathematical school is that it stresses the necessity for mathematicians to learn mathematics as broadly as they can, rather than focusing on a narrow field in order to get important results as soon as possible.

The Moscow mathematical school is particularly strong in the areas of algebra/algebraic geometry, analysis, geometry and topology, probability, mathematical physics and dynamical systems. The scenarios in which these areas were able to develop in Moscow have passed into history. However, it is possible to maintain and develop the Moscow mathematical tradition in new formats, taking into account modern realities such as globalization and mobility of science. There are three recently created centers—the Independent University of Moscow, the Faculty of Mathematics at the National Research University Higher School of Economics (HSE) and the Center for Advanced Studies at Skolkovo Institute of Science and Technology (SkolTech)—whose mission is to strengthen the Moscow mathematical tradition in new ways. HSE and SkolTech are universities offering officially licensed fulltime educational programs. Mathematical curricula at these universities follow not only the Russian and Moscow tradition but also new global developments in mathematics. Mathematical programs at the HSE are influenced by those of the Independent University of Moscow (IUM). The IUM is not a formal university; it is rather a place where mathematics students of different universities can attend special topics courses as well as courses elaborating the core curriculum. The IUM was the main initiator of the HSE Faculty of Mathematics. Nowadays, there is a close collaboration between the two institutions.

While attempting to further elevate traditionally strong aspects of Moscow mathematics, we do not reproduce the former conditions. Instead of isolation and academic inbreeding, we foster global sharing of ideas and international cooperation. An important part of our mission is to make the Moscow tradition of mathematics at a university level a part of global culture and knowledge.

The "Moscow Lectures" series serves this goal. Our authors are mathematicians of different generations. All follow the Moscow mathematical tradition, and all teach or have taught university courses in Moscow. The authors may have taught mathematics at HSE, SkolTech, IUM, the Science and Education Center of the Steklov Institute, as well as traditional schools like MechMath in MGU or MIPT. Teaching and writing styles may be very different. However, all lecture notes are supposed to convey a live dialog between the instructor and the students. Not only personalities of the lecturers are imprinted in these notes, but also those of students.

We hope that expositions published within the "Moscow lectures" series will provide clear understanding of mathematical subjects, useful intuition, and a feeling of life in the Moscow mathematical school.

Moscow, Russia

Igor M. Krichever
Vladlen A. Timorin
Michael A. Tsfasman
Victor A. Vassiliev

Introduction

This book is based on the interrelated courses in complex analysis, the theory of Riemann surfaces, and the theory of integrable systems repeatedly taught by the author at the Independent University of Moscow and at the Department of Mathematics of the Higher School of Economics. The only prerequisite for reading it is a basic knowledge of calculus covered in the first 2 years of undergraduate study (see, e.g., [19]). The theoretical material is complemented by exercises of various degrees of difficulty.

The first two chapters are devoted to the *classical theory of holomorphic and meromorphic functions* and mostly correspond to a standard course in complex analysis. The importance of this theory lies in the fact that the language of holomorphic and meromorphic functions is used to state fundamental laws of physics. The condition for a function to be holomorphic (i.e., to have a complex derivative) turns out to be much more restrictive than the condition to have a real derivative. Holomorphic functions have a number of nice general properties, the most important of which is that the global properties of a function are to a large extent determined by its local properties. This allows one to make important predictions about scientific phenomena relying on local properties of a process.

Then we prove the *classical Riemann theorem*, which says that an arbitrary proper simply connected domain in the complex plane can be mapped onto the standard unit disk by a one-to-one conformal map. Such maps are described by biholomorphic functions. Here, we give the classical proof of the Riemann theorem. Unfortunately, it gives no recipe for constructing the desired map. However, the map itself, which sends an arbitrary domain to the disk, plays a key role in important applications of mathematics (hydromechanics, aerodynamics, and even oil and gas industry [28]). Significant progress in the computation of the desired map was made in this century using the theory of harmonic functions and integrable systems. This new theory is considered in the last chapter of the book.

Chapter 4 is devoted to the *theory of harmonic functions*. It is closely related to the theory of holomorphic functions and is extensively used in various applied problems. The central object of study here is the Green's function of an arbitrary domain and its application to the solution of the Dirichlet problem.

In the subsequent chapters, we turn to Riemann surfaces. A Riemann surface is a one-dimensional complex manifold. Riemann surfaces arise in a natural way and play an important role in various areas of mathematics (the theory of analytic functions, algebraic geometry, spectral theory, the theory of automorphic functions, etc.). Without them, modern mathematics and mathematical physics are unimaginable.

The first of these chapters is devoted to the study of the set of all Riemann surfaces with finitely generated fundamental groups. This set is called the *moduli space of Riemann surfaces*. This was introduced by Riemann, and since then interest in this space only grows. Recently, it turned out, in particular, that the moduli space of Riemann surfaces is related to modern theories of quantum gravity, generates topological invariants of manifolds, etc. The topology of the moduli space is described by the Fricke–Klein theorem. Its original proof (modulo the uniformization theorem) occupies two extensive volumes [5]. In this book, we prove the Fricke–Klein theorem in a single chapter. The proof is based on the study of the geometry of Fuchsian groups [13, 14] and uses only the uniformization theorem. This approach allows one also to study many other moduli spaces related to Riemann surfaces [17].

In the next two chapters, we describe the *classical results of the theory of compact Riemann surfaces*. The first of these chapters is devoted to the general properties of meromorphic functions and differentials on Riemann surfaces. In particular, we prove (modulo the theorem on the existence of holomorphic differentials) that the category of compact Riemann surfaces is isomorphic to the category of complex algebraic curves.

The next chapter is devoted to the proof of the Riemann–Roch theorem and its remarkable corollaries: Weierstrass points, Abel's theorem, etc. Then we turn to the theory of theta functions, including the Jacobi inversion problem. Theta functions of Riemann surfaces were used as early as the nineteenth century to solve complicated differential equations (S. V. Kovalevskaya and others). But a systematic study of these interrelations was initiated only in the late twentieth century in connection with the development of soliton theory [3, 23].

In Chap. 8, the relation between the theory of Riemann surfaces and the theory of differential equations is discussed in detail through the example of the *Kadomtsev–Petviashvili equation*. The Kadomtsev–Petviashvili equation first arose in the description of oscillations in plasma. Later, it turned out that this equation gives a good description of a wide class of wave processes important for applications. It also turned out that the Kadomtsev–Petviashvili equation is the first equation in an infinite system of differential equations for a function of infinitely many variables. This system is called the KP hierarchy [1]. Besides, it turned out that the KP hierarchy arises in a natural way in many areas of mathematics and mathematical physics, from wave theory to topology to algebraic geometry.

We begin Chap. 8 with explicit descriptions, obtained in [16], of the KP hierarchy and its important n-KdV reductions. Then, we describe the theory of Baker–Akhiezer functions suggested by I. M. Krichever. These functions are an analog of exponentials for compact Riemann surfaces of arbitrary genus. Baker–Akhiezer functions allow one, in particular, to find quasi-periodic solutions to the KP equation in the form of expressions involving theta functions [10, 11].

For curves satisfying additional properties, these solutions turn into solutions to the n-KdV hierarchy. The simplest and most important of these solutions are solutions to the Korteweg–de Vries equation, corresponding to hyperelliptic Riemann surfaces. In this case, Krichever's solutions turn into the solutions to KdV found a little earlier by Its and Matveev [7]. A method to effectivize theta-functional solutions to KP was developed by Dubrovin [2].

The last chapter is devoted to the *effectivization of the classical Riemann theorem*, i.e., to an explicit description of a one-to-one biholomorphic function sending an arbitrary simply connected domain with analytic boundary in the plane to the standard unit disk. Progress in this area, which is most important both for theory and applications, has been made fairly recently by Wiegmann and Zabrodin [29], and also in the subsequent papers [9, 12]. The method suggested by them relies on the theory of harmonic functions and the theory of integrable systems. It turned out that the desired biholomorphic functions for all domains can be obtained from partial derivatives of a single function F in infinitely many variables by substituting into it parameters describing the domain. Moreover, this function F is a special solution to an integrable system, already known at the time, called the two-dimensional dispersionless Toda system.

Thus, the effectivization of the Riemann theorem reduces to finding the Taylor series expansion of the function F. This series is of great independent interest also in mathematical physics (e.g., gravitational theory). The first algorithm for computing the Taylor series of the function F was found in [18]. In this book, we follow the approach suggested in [20, 21]. It is based on constructing a general theory of the symmetric dispersionless Toda system, which is of independent interest for the theory of Hurwitz numbers.

I am grateful to A. G. Sergeev and S. N. Malygin for valuable suggestions and for their help in editing the manuscript.

Contents

Chapter 1
Holomorphic Functions

1.1 Complex Derivative

By a *domain* we mean a connected open subset of the complex plane. The correspondence $(x, y) \leftrightarrow z = x + iy$ between the real plane $\mathbb{R}^2$ and the complex plane $\mathbb{C}$ allows one to regard a complex-valued function of a complex variable as

- a map from a domain $D \subset \mathbb{C}$ in the complex plane to the complex plane $\mathbb{C}$ (notation: $w = f(z)$);
- a map from a domain $D \subset \mathbb{R}^2$ in the real plane to the complex plane $\mathbb{C}$ (notation: $w = f(x, y)$);
- a map from a domain $D \subset \mathbb{R}^2$ in the real plane to the real plane $\mathbb{R}^2$ (notation: $(u, v) = f(x, y), u = u(x, y), v = v(x, y)$).

In what follows, we will often switch between these interpretations.

Definition 1.1 Let f be a map from a domain $D \subset \mathbb{C}$ to $\mathbb{C}$ and $z_0 \in D$. If the limit

$$f'(z_0) = \lim_{\Delta z \to 0} \frac{f(z_0 + \Delta z) - f(z_0)}{\Delta z}$$

exists and is finite, then $f'(z_0)$ is called the *complex derivative* of the function f at the point z_0.

Now, let us regard f as a map from a domain in the real plane to the real plane. Then its partial derivative in any direction coincides with $f'(z_0) = f'(x_0, y_0)$. Calculating the partial derivatives in the directions x and y, we obtain

$$f'(z_0) = \frac{\partial f}{\partial x}(z_0)$$

$$= \lim_{\Delta x \to 0} \frac{(u(x_0 + \Delta x, y_0) + iv(x_0 + \Delta x, y_0)) - (u(x_0, y_0) + iv(x_0, y_0))}{\Delta x}$$

© Springer Nature Switzerland AG 2019
S. M. Natanzon, *Complex Analysis, Riemann Surfaces and Integrable Systems*,
Moscow Lectures 3, https://doi.org/10.1007/978-3-030-34640-9_1

$$= \frac{\partial u}{\partial x}(z_0) + i \frac{\partial v}{\partial x}(z_0),$$

$$f'(z_0) = \frac{\partial f}{\partial y}(z_0)$$

$$= \lim_{\Delta x \to 0} \frac{(u(x_0, y_0 + \Delta y) + iv(x_0, y_0 + \Delta y)) - (u(x_0, y_0) + iv(x_0, y_0))}{i \Delta y}$$

$$= -i \frac{\partial u}{\partial y}(z_0) + \frac{\partial v}{\partial y}(z_0).$$

The coincidence of these derivatives implies the following.

Lemma 1.1 *If a function f has a complex derivative at a point z_0, then it satisfies the Cauchy–Riemann equations at this point:*

$$\frac{\partial u}{\partial x}(z_0) = \frac{\partial v}{\partial y}(z_0), \quad \frac{\partial v}{\partial x}(z_0) = -\frac{\partial u}{\partial y}(z_0).$$

Now, we introduce some important notation:

$$\frac{\partial}{\partial z} \equiv \frac{1}{2}\left(\frac{\partial}{\partial x} - i\frac{\partial}{\partial y}\right), \quad \frac{\partial}{\partial \bar{z}} \equiv \frac{1}{2}\left(\frac{\partial}{\partial x} + i\frac{\partial}{\partial y}\right).$$

In this notation,

$$\frac{\partial f}{\partial z} = \frac{\partial(u+iv)}{\partial z} = \frac{\partial u}{\partial z} + i\frac{\partial v}{\partial z} = \frac{1}{2}\left(\frac{\partial u}{\partial x} - i\frac{\partial u}{\partial y}\right) + i\frac{1}{2}\left(\frac{\partial v}{\partial x} - i\frac{\partial v}{\partial y}\right)$$

$$= \frac{1}{2}\left(\frac{\partial u}{\partial x} + i\frac{\partial v}{\partial x}\right) + \frac{1}{2}\left(-i\frac{\partial u}{\partial y} + \frac{\partial v}{\partial y}\right),$$

$$\frac{\partial f}{\partial \bar{z}} = \frac{\partial(u+iv)}{\partial \bar{z}} = \frac{\partial u}{\partial \bar{z}} + i\frac{\partial v}{\partial \bar{z}} = \frac{1}{2}\left(\frac{\partial u}{\partial x} + i\frac{\partial u}{\partial y}\right) + i\frac{1}{2}\left(\frac{\partial v}{\partial x} + i\frac{\partial v}{\partial y}\right)$$

$$= \frac{1}{2}\left(\frac{\partial u}{\partial x} - \frac{\partial v}{\partial y}\right) + i\frac{1}{2}\left(\frac{\partial u}{\partial y} + \frac{\partial v}{\partial x}\right).$$

Thus, we have the following lemma.

Lemma 1.2 *If a function f has a complex derivative at a point z_0, then $\frac{\partial f}{\partial \bar{z}}(z_0) = 0$ and $f'(z_0) = \frac{\partial f}{\partial z}(z_0)$.*

1.2   The Differential of a Complex Function

Lemma 1.3 *If a function* $\mathbb{C} \supset D \xrightarrow{f} \mathbb{C}$ *has a complex derivative at a point* $z_0 = x_0 + i y_0$, *then the corresponding map* $\mathbb{R}^2 \supset D \xrightarrow{f} \mathbb{R}^2$, *regarded as a map from a domain in the real plane to the real plane, is differentiable at the point* (x_0, y_0).

Proof Put

$$\alpha(\Delta z) = \frac{f(z_0 + \Delta z) - f(z_0)}{\Delta z} - f'(z_0).$$

Then

$$
\begin{aligned}
f(z_0 + \Delta z) - f(z_0) &= f'(z_0)\Delta z + \alpha(\Delta z)\Delta z \\
&= \left(\frac{\partial u}{\partial x} + i\frac{\partial v}{\partial x}\right)(\Delta x + i\Delta y) + \alpha(\Delta z)\Delta z \\
&= \left(\frac{\partial u}{\partial x}\Delta x - \frac{\partial v}{\partial x}\Delta y\right) + i\left(\frac{\partial v}{\partial x}\Delta x + \frac{\partial u}{\partial x}\Delta y\right) + \alpha(\Delta z)\Delta z.
\end{aligned}
$$

Therefore,

$$
(u(x_0 + \Delta x, y_0 + \Delta y), v(x_0 + \Delta x, y_0 + \Delta y)) - (u(x_0, y_0), v(x_0, y_0))
$$

$$
= \begin{pmatrix} \dfrac{\partial u}{\partial x} & -\dfrac{\partial v}{\partial y} \\ \dfrac{\partial v}{\partial x} & \dfrac{\partial u}{\partial y} \end{pmatrix} + o(|\Delta z|).
$$

Theorem 1.1 *A function* $\mathbb{C} \supset D \xrightarrow{f} \mathbb{C}$ *has a complex derivative at a point* $z_0 = x_0 + i y_0$ *if and only if the corresponding map* $\mathbb{R}^2 \supset D \xrightarrow{f} \mathbb{R}^2$ *is differentiable at the point* (x_0, y_0) *and satisfies the Riemann–Cauchy equations.*

Proof Assume that a map $\mathbb{R}^2 \supset D \xrightarrow{f} \mathbb{R}^2$ is differentiable at (x_0, y_0). Then

$$
u(x_0 + \Delta x, y_0 + \Delta y) - u(x_0, y_0) = \left(\frac{\partial u}{\partial x}\Delta x + \frac{\partial u}{\partial y}\Delta y\right) + o(|\Delta z|),
$$

$$
v(x_0 + \Delta x, y_0 + \Delta y)) - v(x_0, y_0) = \left(\frac{\partial v}{\partial x}\Delta x + \frac{\partial v}{\partial y}\Delta y\right) + o(|\Delta z|).
$$

Hence for $\mathbb{R}^2 \supset D \xrightarrow{f} \mathbb{C}$ we have

$$
\begin{aligned}
f(z_0 + \Delta z) - f(z_0) &= \left(\frac{\partial u}{\partial x}\Delta x + \frac{\partial u}{\partial y}\Delta y\right) + i\left(\frac{\partial v}{\partial x}\Delta x + \frac{\partial v}{\partial y}\Delta y\right) + o(|\Delta z|) \\
&= \frac{1}{2}\left[\left(\frac{\partial}{\partial x} - i\frac{\partial}{\partial y}\right)(u + iv)\right](\Delta x + i\,\Delta y) \\
&\quad + \frac{1}{2}\left[\left(\frac{\partial}{\partial x} + i\frac{\partial}{\partial y}\right)(u + iv)\right](\Delta x - i\,\Delta y) + o(|\Delta z|) \\
&= \frac{\partial f}{\partial z}\Delta z + \frac{\partial f}{\partial \bar{z}}(\Delta x - i\,\Delta y) + o(|\Delta z|).
\end{aligned}
$$

If f satisfies the Riemann–Cauchy equations, then the calculations from the previous section show that $\frac{\partial f}{\partial \bar{z}} = 0$, whence

$$
f(z_0 + \Delta z) - f(z_0) = \frac{\partial f}{\partial z}\Delta z + o(|\Delta z|),
$$

i.e., f has a complex derivative at z_0.

The converse statement follows from Lemmas 1.1 and 1.3.

Theorem 1.2 *Assume that functions* $\mathbb{C} \supset D \xrightarrow{f,g} \mathbb{C}$ *have complex derivatives at a point* z_0. *Then the functions* $f \pm g$, $f \cdot g$, *and* $\frac{f}{g}$ *(if* $g(z_0) \neq 0$) *have complex derivatives at the point* z_0, *and*

$$
(f \pm g)'(z_0) = f'(z_0) \pm g'(z_0),
$$
$$
(fg)'(z_0) = f'(z_0)g(z_0) + f(z_0)g'(z_0),
$$
$$
\left(\frac{f}{g}\right)'(z_0) = \frac{f'(z_0)g(z_0) - f(z_0)g'(z_0)}{g^2(z_0)}.
$$

Theorem 1.3 *Let* $\mathbb{C} \supset D \xrightarrow{f} V \xrightarrow{g} \mathbb{C}$, $V \subset \mathbb{C}$, *and assume that* f *has a complex derivative at a point* z_0 *and* g *has a complex derivative at the point* $f(z_0)$. *Then the function* $\varphi(z) = g(f(z))$ *has a complex derivative at the point* z_0 *and* $\varphi'(z_0) = g'(f(z_0))f'(z_0)$.

Exercise 1.1 Prove Theorems 1.2 and 1.3. (*Hint.* These proofs follow along the same lines as the proofs of the corresponding theorems of real analysis.)

1.3   Holomorphic Functions

Definition 1.2 A function $D \xrightarrow{f} \mathbb{C}$ is said to be *holomorphic on the domain* $D \subset \mathbb{C}$ if it has a complex derivative at every point $z \in D$, and *holomorphic at a point* z_0 if it is holomorphic in some neighborhood $V \ni z_0$ of z_0.

Examples of holomorphic functions:

1. $f(z) = \text{const}$, $f'(z) = 0$;
2. $f(z) = az$ with $a \neq 0$; if $a = re^{i\varphi}$, then f rotates the plane $\mathbb{C}$ about 0 by the angle φ and stretches or shrinks this plane by the factor r;
3. $f(z) = z^n$; the function f increases the angle between rays starting at 0 by the factor n.

Exercise 1.2 Show that if $f'(z) = 0$ on the whole domain $D \subset \mathbb{C}$, then $f = \text{const}$ on D.

If $f'(z_0) \neq 0$, then in a small neighborhood of z_0 the action of f is almost the same as in Example 2. More exactly, the following result holds.

Exercise 1.3 Let $f'(z_0) \neq 0$. Show that the function f preserves the angle between curves intersecting at z_0.

Definition 1.3 An angle-preserving map is said to be *conformal*.

As in the real case, we can consider sequences and series of complex functions $f(z) = \sum_{n=1}^{\infty} f_n(z)$. All definitions and theorems carry over literally to the complex case if instead of intervals $\{x \in \mathbb{R} \mid |x - x_0| < r\}$ one considers disks $\{z \in \mathbb{C} \mid |z - z_0| < r\}$.

Exercise 1.4 Show that if a series $f(z) = \sum_{n=0}^{\infty} a_n(z)$ converges at least at one point of a domain $D \subset \mathbb{C}$ and the series $\sum_{n=0}^{\infty} a_n'(z)$ converges uniformly on D, then the series $f(z)$ converges uniformly on D and $f'(z) = \sum_{n=0}^{\infty} a_n'(z)$.

We will be mainly interested in power series

$$f(z) = \sum_{n=0}^{\infty} c_n (z - z_0)^n, \quad c_n \in \mathbb{C}.$$

Exercise 1.5 Let $\frac{1}{R} = \overline{\lim\limits_{n \to \infty}} \sqrt[n]{|c_n|}$. Show that the power series $f(z)$ converges absolutely on $D = \{z \in \mathbb{C} \mid |z - z_0| < R\}$, diverges on $\mathbb{C} \setminus D = \{z \in \mathbb{C} \mid |z - z_0| > R\}$, and converges uniformly on every compact subset $K \subset D$.

Set

$$e^z = \exp z = \sum_{n=0}^{\infty} \frac{z^n}{n!},$$

$$\cos z = \sum_{k=0}^{\infty}(-1)^k \frac{z^{2k}}{(2k)!} = \frac{1}{2}(e^{iz} + e^{-iz}),$$

$$\sin z = \sum_{k=0}^{\infty}(-1)^k \frac{z^{2k+1}}{(2k+1)!} = \frac{1}{2i}(e^{iz} - e^{-iz}).$$

Exercise 1.6 Show that the functions e^z, $\cos z$, $\sin z$ exist and are holomorphic in the whole plane $\mathbb{C}$. For each of these functions, find a domain D such that $f(D) = \mathbb{C}$.

1.4 Complex Integration

By a *curve* (or *path*) we mean the oriented image of a piecewise smooth map from an interval $[\alpha, \beta]$ to the plane $\mathbb{C}$. A closed curve is called a *contour*. In the real case, the integral of a function f over a curve γ is the limit, as $\max |\Delta z_k| \to 0$, of the Riemann sums $S = \sum_{k=0}^{n} f(\xi_k)\Delta z_k$ where ξ_k are points on γ and Δz_k are the distances between these points. If we formally take the variable and the values of f to be complex numbers, we obtain the complex integral of f over the curve γ in $\mathbb{C}$. Here,

$$S = \sum_{k=0}^{n}\big(u(\xi_k) + iv(\xi_k)\big)(\Delta x + i\Delta y)$$

$$= \sum_{k=0}^{n}\big(u(\xi_k)\Delta x_k - v(\xi_k)\Delta y_k\big) + i\big(u(\xi_k)\Delta y_k + v(\xi_k)\Delta x_k\big).$$

Thus, we arrive at the following definition.

Definition 1.4 The *integral of a function* $f(z) = u(x, y) + iv(x, y)$ over a curve γ in $\mathbb{C}$ is the complex number

$$\int_{\gamma} f(z)\,dz \overset{\text{def}}{=} \int_{\gamma}(u\,dx - v\,dy) + i\int_{\gamma}(u\,dy + v\,dx).$$

If $w : [\alpha, \beta] \to \mathbb{C}$ is a smooth parametrization of the curve γ and $w(t) = x(t) + iy(t)$, then

$$\int_\gamma f(z)\,dz = \int_\alpha^\beta \left(u(w(t))x'(t)\,dt - v(w(t))y'(t)\,dt \right)$$

$$+ i \int_\alpha^\beta \left(u(w(t))y'(t)\,dt + v(w(t))x'(t)\,dt \right) = \int_\alpha^\beta f(w(t))w'(t)\,dt.$$

In particular, the right-hand side does not depend on the parametrization $w(t)$.

Example 1.1 Let $\gamma = \{z \in \mathbb{C} \mid |z - a| = r\} = \{a + re^{it} \mid t \in [0, 2\pi]\}$. Then

$$\int_\gamma (z - a)^n\,dz = r^{n+1}i \int_0^{2\pi} e^{i(n+1)t}\,dt = \begin{cases} 0 & \text{if } n \neq -1, \\ 2\pi i & \text{if } n = -1. \end{cases}$$

Example 1.2 Let $n \neq -1$ and γ be a path in $\mathbb{C}$ from a to b. Consider a parametrization $w = w(t)$ of γ. Then

$$\int_\gamma z^n\,dz = \int_\alpha^\beta w^n(t)w'(t)\,dt = \frac{1}{n+1} \int_\alpha^\beta \left(\frac{d}{dt}(w^{n+1}(t)) \right) dt$$

$$= \frac{1}{n+1}(w^{n+1}(\beta) - w^{n+1}(\alpha)) = \frac{b^{n+1} - a^{n+1}}{n+1}.$$

By $\int_\gamma |f|\,|dz|$ we will denote the arc-length integral $\int_\alpha^\beta |f|\,|z'(t)|\,dt$.
The following theorem is obvious from the definition.

Theorem 1.4

1. *Reversing the orientation of γ changes the sign of the integral.*
2. *The following equalities hold:*

$$\int_\gamma (af + bg)\,dz = a \int_\gamma f\,dz + b \int_\gamma g\,dz, \quad \int_{\gamma_1 \cup \gamma_2} f\,dz = \int_{\gamma_1} f\,dz + \int_{\gamma_2} f\,dz.$$

3. *The following inequality holds:* $|\int_\gamma f\,dz| \leq \int_\gamma |f|\,|dz|$.

   In particular, if $|f(z)| \leq M$, then $\left| \int_\gamma f\,dz \right| \leq M|\gamma|$ where $|\gamma|$ is the length of γ.

1.5   Cauchy's Theorem

In what follows, we assume that domains in the complex plane have the standard orientation (the counterclockwise direction is considered positive). The orientation of a domain determines an orientation of its boundary.

Lemma 1.4 *Let $f(z)$ be a holomorphic function on a domain D. Then $\int_{\partial \Delta} f(z)\, dz = 0$ for every oriented triangle $\Delta \subset D$.*

Proof Let $|\int_{\partial \Delta} f(z)\, dz| = M > 0$. Divide the triangle Δ into four triangles a_1, a_2, a_3, a_4 as shown in Fig. 1.1. Then $M = \left| \sum_{i=1}^{n} \int_{\partial a_i} f(z)\, dz \right| \le \sum_{i=1}^{n} \left| \int_{\partial a_i} f(z)\, dz \right|$.

Hence, for one of the triangles $\Delta_1 \in \{a_1, a_2, a_3, a_4\}$ we have $\left| \int_{\partial \Delta_1} f\, dz \right| \ge \frac{1}{4} M$.

Continuing in the same way, we get a sequence of triangles $\Delta \supset \Delta_1 \supset \Delta_2 \supset \dots$ such that $|\int_{\partial \Delta_n} f(z)\, dz| \ge \frac{1}{4^n} M$.

Let $z_0 \in \bigcap \Delta_i \subset D$. Put $\alpha(z) = \frac{f(z) - f(z_0)}{z - z_0} - f'(z_0)$. Then for every $\varepsilon > 0$ there exists $\delta > 0$ such that $|a(z)| < \varepsilon$ for $0 < |z - z_0| < \delta$. Let $\Delta_n \subset \{z \in \mathbb{C} \mid |z - z_0| < \delta\}$; then, by Example 1.1, we have

$$\left| \int_{\partial \Delta_n} f(z)\, dz \right| = \left| \int_{\partial \Delta_n} f(z_0)\, dz + \int_{\partial \Delta_n} f'(z_0)(z - z_0)\, dz + \int_{\partial \Delta_n} \alpha(z)(z - z_0)\, dz \right|$$

$$= \left| \int_{\partial \Delta_n} \alpha(z)(z - z_0)\, dz \right| \le \int_{\partial \Delta_n} |\alpha(z)|\, |z - z_0|\, |dz| \le \varepsilon |\partial \Delta_n|^2$$

$$= \varepsilon \left(\frac{|\partial \Delta|}{2^n} \right)^2 = \varepsilon \frac{|\partial \Delta|^2}{4^n}.$$

Thus, $\frac{1}{4^n} M \le \left| \int_{\partial \Delta_n} f(z)\, dz \right| \le \varepsilon \frac{|\partial \Delta|^2}{4^n}$, whence $M \le \varepsilon |\partial \Delta|^2$ and, therefore, $M = 0$.

Fig. 1.1

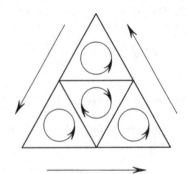

Fig. 1.2

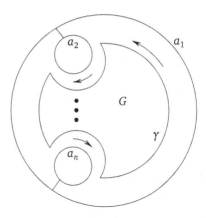

Theorem 1.5 (Cauchy) *Let $f(z)$ be a holomorphic function on a domain D and γ be a null-homotopic closed path in D. Then $\int_{\gamma} f(z)\,dz = 0$.*

Proof Let Q be the domain bounded by γ. If γ is a polygonal path, then Q can be divided into finitely many triangles $\Delta_i \subset D$. By Lemma 1.4, we have

$$\int_{\gamma} f(z)\,dz = \sum_{i=1}^{n} \int_{\Delta_i} f(z)\,dz = 0.$$

The integral over an arbitrary path is the limit of integrals over polygonal paths.

Remark 1.1 For functions f having continuous derivatives $f'(z)$, Cauchy's theorem can be derived from Green's theorem:

$$\int_{\gamma} f(z)\,dz = \int_{\partial Q} (u\,dx - v\,dy) + i \int_{\partial Q} (u\,dy + v\,dx)$$

$$= \iint_{Q} \left(-\frac{\partial v}{\partial x} - \frac{\partial u}{\partial y}\right) dx\,dy + i \iint_{Q} \left(\frac{\partial u}{\partial x} - \frac{\partial v}{\partial y}\right) dx\,dy = 0.$$

However, such a proof cannot be used in a systematic treatment of complex analysis, since Cauchy's theorem is applied later to prove that the derivative of every holomorphic function is continuous.

Theorem 1.6 *Let f be a holomorphic function on a domain D and $G \subset D$ be a compact subset bounded by finitely many closed contours. Then $\int_{\partial G} f(z)\,dz = 0$.*

Proof Let us join the boundary components of G by line segments $\delta_1, \ldots, \delta_m \subset G$ in such a way that the set $\tilde{G} = G \setminus \bigcup_{i=1}^{m} \delta_i$ is simply connected (see Fig. 1.2). Then, by Theorem 1.5, we have

$$0 = \int_{\partial \tilde{G}} f(z)\, dz = \int_{\partial G} f(z)\, dz.$$

1.6 Antiderivative

Definition 1.5 An *antiderivative* of a function $f(z)$ on a domain D is a function $F(z)$ holomorphic on D such that $F'(z) = f(z)$.

Exercise 1.7 Let F be an antiderivative of f. Show that G is an antiderivative of f if and only if $G = F + \text{const}$.

Lemma 1.5 *Let $f(z)$ be a holomorphic function on the disk $D = \{z \in \mathbb{C} \mid |z - a| < r\}$. Then $F(z) = \int\limits_{[a,z]} f(w)\, dw$ is an antiderivative of f on D.*

Proof Let $z + h \in D$. Then, by Lemma 1.4 and Example 1.2, we have

$$F(z+h) - F(z) = \int_{[a,z+h]} f(w)\, dw - \int_{[a,z]} f(w)\, dw = \int_{[z,z+h]} f(w)\, dw$$

$$= \int_{[z,z+h]} f(z)\, dw + \int_{[z,z+h]} \big(f(w) - f(z)\big)\, dw = f(z)h + \int_{[z,z+h]} \big(f(w) - f(z)\big)\, dw.$$

Thus,

$$\left| \frac{F(z+h) - F(z)}{h} - f(z) \right| \le \frac{1}{|h|} \int_{[z,z+h]} |\alpha(h)|\, |dh|$$

where $\alpha(h) = f(z+h) - f(z)$. Since f is continuous, for every $\varepsilon > 0$ there exists $\delta > 0$ such that $|\alpha(h)| < \varepsilon$ for $|h| < \delta$. Thus, $\left| \frac{F(z+h)-F(z)}{h} - f(z) \right| \le \frac{1}{|h|} \varepsilon |h| = \varepsilon$ for $|h| < \delta$, i.e., $F'(z) = f(z)$.

Definition 1.6 Let f be a holomorphic function on a domain D. An *antiderivative of f along a curve γ in D* is a continuous function $\varphi(z)$ on γ that is the restriction to γ of an antiderivative of f on a domain $U \subset D$ containing γ.

Theorem 1.7 *Let f be a holomorphic function on a domain D and γ be a non-self-intersecting curve in D that begins at a and ends at b. Then f has an* *antiderivative $\varphi(z)$ along γ, and*

$$\varphi(b) - \varphi(a) = \int_\gamma f(z)\, dz.$$

Proof The curve γ is the preimage of the interval $[0, 1]$ under a continuous function $z: [0, 1] \rightarrow D$. By the uniform continuity of z, the interval $[0, 1]$ can be covered by intervals $\alpha_1, \ldots, \alpha_n$ so that the image $z(\alpha_i)$ is contained in a disk $D_i \subset D$ where $D_i \cap D_j \neq \varnothing$ if and only if $|i - j| = 1$. Using Lemma 1.5, choose an antiderivative $\tilde{\varphi}_i(z)$ on each disk D_i. By Exercise 1.7, these antiderivatives can be chosen so as to coincide on all intersections of the disks. Then we obtain a desired antiderivative φ on the union U of these disks. The equality $\varphi(b) - \varphi(a) = \int_\gamma f(z)\, dz$ follows from the explicit construction of an antiderivative used in Lemma 1.5.

Theorem 1.8 *If f is a holomorphic function on a connected, simply connected domain D, then f has an antiderivative F on D, and $\int_\gamma f(z)\, dz = F(b) - F(a)$ for every path γ in D that begins at a and ends at b.*

Proof Let $a \in D$. For $z \in D$, set $F(z) = \int_{\gamma_z} f(w)\, dw$ where γ_z is a path in D that connects the points a and z. By Theorem 1.5, this integral does not depend on the choice of γ, and hence F is well defined. By Lemma 1.5 and Theorem 1.7, we have $F'(z) = f(z)$ and $\int_\gamma f(z)\, dz = F(b) - F(a)$.

1.7 Cauchy's Integral Formula

Theorem 1.9 (Mean Value Theorem) *Let f be a holomorphic function on a domain D and $G = \{z \in \mathbb{C} \mid |z - z_0| \leq r\} \subset D$. Then*

$$f(z_0) = \frac{1}{2\pi i} \int_{\partial G} \frac{f(z)}{z - z_0}\, dz = \frac{1}{2\pi} \int_0^{2\pi} f(z_0 + re^{it})\, dt.$$

Proof Since f is continuous, for every $\varepsilon > 0$ there exists ρ such that $r > \rho > 0$ and $|f(z) - f(z_0)| < \varepsilon$ for $|z - z_0| \leq \rho$. Set $G_\rho = \{z \in \mathbb{C} \mid |z - z_0| \leq \rho\}$. By Theorem 1.6, we have

$$0 = \int_{\partial G} \frac{f(z)}{z - z_0}\, dz - \int_{\partial G_\rho} \frac{f(z)}{z - z_0}\, dz.$$

Hence, by Example 1.1, we obtain

$$f(z_0) - \frac{1}{2\pi i} \int\limits_{\partial G} \frac{f(z)}{z - z_0} \, dz = f(z_0) - \frac{1}{2\pi i} \int\limits_{\partial G_\rho} \frac{f(z)}{z - z_0} \, dz$$

$$= f(z_0) \frac{1}{2\pi i} \int\limits_{\partial G_\rho} \frac{dz}{z - z_0} - \frac{1}{2\pi i} \int\limits_{\partial G_\rho} \frac{f(z)}{z - z_0} \, dz = \frac{1}{2\pi i} \int\limits_{\partial G_\rho} \frac{f(z) - f(z_0)}{z - z_0} \, dz.$$

It follows that

$$\left| f(z_0) - \frac{1}{2\pi i} \int\limits_{\partial G} \frac{f(z)}{z - z_0} \, dz \right| \le \frac{1}{2\pi} \int\limits_{\partial G_\rho} \frac{\varepsilon}{\rho} |dz| = \varepsilon,$$

which, since ε is arbitrary, implies that

$$f(z_0) = \frac{1}{2\pi i} \int\limits_{\partial G} \frac{f(z)}{z - z_0} \, dz.$$

Substituting $z = z_0 + re^{it}$, we see that

$$f(z_0) = \frac{1}{2\pi i} \int\limits_{\partial G} \frac{f(z)}{z - z_0} \, dz = \frac{1}{2\pi i} \int\limits_0^{2\pi} \frac{f(z_0 + re^{it})}{re^{it}} \cdot (z_0 + re^{it})' \, dt$$

$$= \frac{1}{2\pi} \int\limits_0^{2\pi} f(z_0 + re^{it}) \, dt.$$

Theorem 1.10 (Cauchy's Integral Formula) *Let f be a holomorphic function on a domain D and $G \subset D$ be a compact set bounded by finitely many contours. Then*

$$\frac{1}{2\pi i} \int\limits_{\partial G} \frac{f(z)}{z - z_0} \, dz = \begin{cases} f(z_0) & \text{if } z_0 \in G \setminus \partial G, \\ 0 & \text{if } z_0 \notin G. \end{cases}$$

Proof If $z_0 \notin G$, then the claim follows from Theorem 1.6. If $z_0 \in G \setminus \partial G$, then consider $U = \{z \in \mathbb{C} \mid |z - z_0| \le r\} \subset G \setminus \partial G$. By Theorems 1.9 and 1.6, we have

$$f(z_0) = \frac{1}{2\pi i} \int\limits_{\partial U} \frac{f(z)}{z - z_0} \, dz$$

$$= \frac{1}{2\pi i} \int\limits_{\partial U} \frac{f(z)}{z - z_0} \, dz + \frac{1}{2\pi i} \int\limits_{\partial (G \setminus U)} \frac{f(z)}{z - z_0} \, dz = \frac{1}{2\pi i} \int\limits_{\partial G} \frac{f(z)}{z - z_0} \, dz.$$

1.8   Taylor Series Expansion

Theorem 1.11 *Let f be a holomorphic function on a domain D and*

$$G = \{z \in \mathbb{C} \mid |z - z_0| < R\} \subset D.$$

Then f coincides on G with the sum of the series

$$\sum_{n=0}^{\infty} c_n (z - z_0)^n$$

where $c_n = \frac{1}{2\pi i} \int_{\gamma} \frac{f(z)}{(z-z_0)^{n+1}} \, dz$ and $\gamma = \{z \in \mathbb{C} \mid |z - z_0| = r < R\}$.

Proof By Theorem 1.6, the coefficients c_n do not depend on the choice of $r < R$. Let $z \in G$ and $|z - z_0| < r$. Consider the series

$$\frac{1}{w - z} = \frac{1}{(w - z_0)\left(1 - \frac{z-z_0}{w-z_0}\right)} = \sum_{n=0}^{\infty} \frac{(z - z_0)^n}{(w - z_0)^{n+1}}.$$

If $w \in \gamma$, then

$$\left| \frac{(z - z_0)^n}{(w - z_0)^{n+1}} \right| = \frac{1}{|z - z_0|} \left(\left| \frac{z - z_0}{w - z_0} \right| \right)^{n+1} = \frac{\rho^{n+1}}{|z - z_0|}$$

where $\rho = \frac{|z-z_0|}{r} < 1$. Thus, the series $\sum_{n=0}^{\infty} \frac{(z-z_0)^n}{(w-z_0)^{n+1}}$ is dominated by an absolutely convergent series and, therefore, converges uniformly with respect to w on γ. The function $f(w)$ is bounded on γ, hence the series $\sum_{n=0}^{\infty} f(w) \frac{(z-z_0)^n}{(w-z_0)^{n+1}}$ also converges uniformly with respect to w on γ. In particular, it can be integrated termwise, and we obtain, using Cauchy's formula,

$$f(z) = \frac{1}{2\pi i} \int_{\gamma} \frac{f(w)}{w - z} \, dw = \frac{1}{2\pi i} \int_{\gamma} \sum_{n=0}^{\infty} f(w) \frac{(z - z_0)^n}{(w - z_0)^{n+1}} \, dw$$

$$= \sum_{n=0}^{\infty} (z - z_0)^n \int_{\gamma} \frac{f(w)}{2\pi i} \frac{dw}{(w - z_0)^{n+1}} = \sum_{n=0}^{\infty} c_n (z - z_0)^n.$$

Theorem 1.12 *Let* $\frac{1}{R} = \varlimsup\limits_{n \to \infty} \sqrt[n]{|c_n|} < \infty$. *Then the function*

$$f(z) = \sum_{n=0}^{\infty} c_n (z - z_0)^n$$

exists and is holomorphic on the disk $D = \{z \in \mathbb{C} \mid |z - z_0| < R\}$; *the function* $f'(z)$ *is also holomorphic on* D.

Proof Set

$$\varphi(z) = \sum_{n=1}^{\infty} n c_n (z - z_0)^{n-1} = \sum_{n=0}^{\infty} d_n (z - z_0)^n.$$

Since

$$\varlimsup_{n \to \infty} \sqrt[n]{|d_n|} = \varlimsup_{n \to \infty} \sqrt[n]{n |c_n|} = \varlimsup_{n \to \infty} \sqrt[n]{|c_n|} = \frac{1}{R},$$

the function $\varphi(z)$ is defined on D and converges uniformly on compact subsets of D. Hence, $\varphi(z)$ can be integrated termwise over paths in D. Set

$$F(z) = \int_{[z_0, z]} \varphi(w)\, dw = \int_{[z_0, z]} \sum_{n=1}^{\infty} n c_n (w - z_0)^{n-1} dw$$

$$= \sum_{n=1}^{\infty} n c_n \int_{[z, z_0]} (w - z_0)^{n-1} dw = \sum_{n=1}^{\infty} n c_n \frac{1}{n} (w - z_0)^n \big|_{w=z_0}^{w=z} = f(z) - c_0.$$

On the other hand,

$$F(z + h) - F(z) = \int_{[z, z+h]} \varphi(w)\, dw$$

$$= \sum_{n=1}^{\infty} n c_n \int_{[z, z+h]} (w - z_0)^{n-1} dw = \sum_{n=1}^{\infty} c_n (w - z_0)^n \big|_z^{z+h}$$

$$= \sum_{n=1}^{\infty} c_n \big[(z + h - z_0)^n - (z - z_0)^n \big] = h \sum_{n=1}^{\infty} n c_n (z - z_0)^{n-1} + h^2 g(z).$$

Thus, the function

$$f'(z) = \lim_{h \to 0} \frac{f(z+h) - f(z)}{h} = \lim_{h \to 0} \frac{F(z+h) - F(z)}{h} = \sum_{n=1}^{\infty} nc_n(z - z_0)^{n-1} = \varphi(z)$$

is defined on D.

Now, recall that

$$\varphi(z) = \sum_{n=0}^{\infty} d_n(z - z_0)^n,$$

where $\varlimsup\limits_{n \to \infty} \sqrt[n]{|d_n|} = \frac{1}{R}$. This allows us to apply to φ the same argument as used in the analysis of the function f. It shows that the function $\varphi' = f''$ is defined on D and, therefore, the function $f'(z)$ is holomorphic on D.

1.9   A Criterion for a Function to Be Holomorphic

Theorem 1.13 *Let f be a holomorphic function on a domain D. Then on D it has complex derivatives of all orders, they are holomorphic, and*

$$f^{(n)}(z) = \frac{n!}{2\pi i} \int_{\partial U} \frac{f(w)}{(w - z)^{n+1}} \, dw \quad \text{where } U = \{w \in \mathbb{C} \mid |w - z| \le r\} \subset D.$$

Proof Let $z_0 \in D$ and $G = \{z \in \mathbb{C} \mid |z - z_0| \le R\} \subset D$. By Theorem 1.11, the function $f(z)$ can be represented on G as a power series $f(z) = \sum_{n=0}^{\infty} c_n(z - z_0)^n$. Therefore, by Theorem 1.12, the function $f'(z)$ is holomorphic on $G \setminus \partial G$. Repeating this argument, we prove that $f^{(n)}(z)$ is holomorphic for every n. Repeating an argument from real analysis, we find, by termwise differentiation, that $c_n = \frac{1}{n!} f^{(n)}(z_0)$. Comparing with Theorem 1.11, we obtain the desired formulas for $f^{(n)}(z)$.

Theorem 1.14 *Let $U = \{z \in \mathbb{C} \mid |z - a| < r\}$ and $f : U \to \mathbb{C}$. Then the following three conditions are equivalent:*

(1) *the function f is holomorphic on U, i.e., has a complex derivative at every point of U;*

(2) *the function f is continuous on U, and the integral of f over the boundary of every triangle $\Delta \subset U$ vanishes;*

(3) $f(z) = \sum_{n=0}^{\infty} c_n(z - a)^n$ *on U.*

Proof (1) $\Rightarrow$ (2) is Theorem 1.5, (1) $\Rightarrow$ (3) is Theorem 1.11, (3) $\Rightarrow$ (1) is Theorem 1.12. Let us prove that (2) $\Rightarrow$ (1) (Morera's theorem). Put $F(z) = \int_{[a,z]} f(w)\,dw$. Then

$$F(z+h) - F(z) = \int_{[z,z+h]} f(w)\,dw$$

and

$$\left| \frac{F(z+h) - F(z)}{h} - f(z) \right|$$

$$= \frac{1}{|h|}\left| \int_{[z,z+h]} f(w)\,dw - hf(z) \right| \le \frac{1}{|h|}\left| \int_{[z,z+h]} (f(w) - f(z))\,dw \right|$$

$$\le \frac{1}{|h|} \max_{[z,z+h]} |f(w) - f(z)| \cdot |h| = \max_{[z,z+h]} |f(w) - f(z)|.$$

Hence, since the function f is continuous at z, the function F is differentiable at z and $F'(z) = f(z)$. Therefore, the function $F(z)$ is holomorphic on U. By Theorem 1.13, this implies that the function $f(z)$ is holomorphic.

Thus, in contrast to smooth functions of real variable, a holomorphic function is determined by a countable set of numbers. These numbers are determined by the behavior of the function in a neighborhood of a point, so the behavior of the function in a neighborhood of a point determines the whole function. Moreover, these numbers can be found by contour integration, which is sometimes more convenient than differentiation.

1.10 Weierstrass' Theorem

Theorem 1.15 (Weierstrass) *Let $f_n(z)$ be holomorphic functions on a domain D, and assume that the series $f(z) = \sum_{n=0}^{\infty} f_n(z)$ converges uniformly on every compact subset of D. Then the function f is holomorphic and $f'(z) = \sum_{n=0}^{\infty} f_n'(z)$.*

Proof For an arbitrary point $a \in D$, consider the closed disk

$$U = \{z \in \mathbb{C} \mid |z - a| \le R\} \subset D.$$

If $\gamma \subset U$ is the boundary of a triangle, then, by uniform convergence, we have $\int_{\gamma} f(z)\,dz = \sum_{n=0}^{\infty} \int_{\gamma} f_n(z)\,dz$, and, by Theorem 1.14, the function $f(z)$ is holomorphic on U. Besides, by Theorem 1.13, we have

$$f'(a) = \frac{1}{2\pi i} \int_{\partial U} \frac{f(z)}{(z-a)^2} \, dz = \frac{1}{2\pi i} \int_{\partial U} \sum_{n=0}^{\infty} \frac{f_n(z)}{(z-a)^2} \, dz$$

$$= \sum_{n=0}^{\infty} \frac{1}{2\pi i} \int_{\partial U} \frac{f_n(z)}{(z-a)^2} \, dz = \sum_{n=0}^{\infty} f_n'(a).$$

Thus, in contrast to smooth functions of real variable, the set of functions holomorphic on a given domain D is closed with respect to the topology of uniform convergence on compact subsets of D.

Chapter 2
Meromorphic Functions

2.1 Functions Holomorphic on a Ring: Laurent Series

Now we turn to studying the properties of functions holomorphic in non-simply connected domains.

Theorem 2.1 *Let $f(z)$ be a function holomorphic on an annulus*

$$V = \{0 \leq r < |z - a| < R \leq \infty\}.$$

Then on V it can be expanded as

$$f(z) = \sum_{n=-\infty}^{\infty} c_n (z - a)^n$$

where

$$c_n = \frac{1}{2\pi i} \int_{\gamma_\rho} \frac{f(w)}{(w - a)^{n+1}} \, dw \quad and \quad \gamma_\rho = \{z \in \mathbb{C} \mid |z - a| = \rho\} \subset V.$$

Proof Let $z \in V$ and $U = \{w \in V \mid \alpha \leq |w - a| \leq \beta\} \ni z$. By Cauchy's formula,

$$f(z) = \frac{1}{2\pi i} \int_{\partial U} \frac{f(w)}{(w - z)} \, dw = f_\beta(z) - f_\alpha(z)$$

© Springer Nature Switzerland AG 2019
S. M. Natanzon, *Complex Analysis, Riemann Surfaces and Integrable Systems*,
Moscow Lectures 3, https://doi.org/10.1007/978-3-030-34640-9_2

where

$$f_\beta(z) = \frac{1}{2\pi i} \int\limits_{\gamma_\beta} \frac{f(w)}{w - z}\, dw = \frac{1}{2\pi i} \int\limits_{\gamma_\beta} f(w) \frac{1}{(w - a)\left(1 - \frac{z - a}{w - a}\right)}\, dw$$

$$= \frac{1}{2\pi i} \int\limits_{\gamma_\beta} f(w) \sum_{n=0}^{\infty} \frac{(z - a)^n}{(w - a)^{n+1}}\, dw = \sum_{n=0}^{\infty} c_n (z - a)^n,$$

$$f_\alpha(z) = \frac{1}{2\pi i} \int\limits_{\gamma_\alpha} \frac{f(w)}{w - z}\, dw$$

$$= -\frac{1}{2\pi i} \int\limits_{\gamma_\alpha} \frac{f(w)}{(z - a)\left(1 - \frac{w - a}{z - a}\right)}\, dw = -\int\limits_{\gamma_\alpha} \frac{f(w)}{2\pi i} \sum_{n=0}^{\infty} \frac{(w - a)^n}{(z - a)^{n+1}}\, dw$$

$$= -\int\limits_{\gamma_\alpha} \frac{f(w)}{2\pi i} \sum_{n=0}^{\infty} \frac{(z - a)^{-(n+1)}}{(w - a)^{-n}}\, dw = -\sum_{n=1}^{\infty} c_{-n} (z - a)^{-n}.$$

Definition 2.1 A series of the form

$$\sum_{n=-\infty}^{\infty} c_n (z - a)^n$$

is called a *Laurent series with regular part* $\sigma_1 = \sum\limits_{n=0}^{\infty} c_n (z - a)^n$ and *principal part* $\sigma_2 = \sum\limits_{n=-\infty}^{-1} c_n (z - a)^n$.

Theorem 2.2 A *Laurent series* $\sum\limits_{n=-\infty}^{\infty} c_n (z - a)^n$ *defines a function holomorphic on the annulus* $V = \{z \in \mathbb{C} \mid r < |z - a| < R\}$ *where* $r = \varlimsup\limits_{n \to +\infty} \sqrt[n]{|c_{-n}|}$ *and* $R = \frac{1}{\varlimsup\limits_{n \to +\infty} \sqrt[n]{|c_n|}}$.

Proof By Abel's theorem, the functions σ_1 and σ_2 converge uniformly on every compact subset of V. Therefore, by Weierstrass' theorem, these functions are holomorphic on the annulus V.

Theorem 2.3 (Cauchy's Inequality) Let $f(z) = \sum\limits_{n=-\infty}^{\infty} c_n (z - a)^n$ be a function holomorphic on an annulus $V = \{z \mid r < |z - a| < R\}$ and $\gamma_\rho = \{z \mid |z - a| = \rho\} \subset V$. Then

$$c_n = \frac{1}{2\pi i} \int\limits_{\gamma_\rho} \frac{f(w)}{(w - a)^{n+1}}\, dw \quad and \quad |c_n| \le \frac{M}{\rho^n} \quad where \; M = \max\limits_{\gamma_\rho} |f|.$$

Proof By Example 1.1, we have

$$\int_{\gamma_\rho} \frac{f(z)\,dz}{(z-a)^{n+1}} = \int_{\gamma_\rho} \sum_{m=-\infty}^{\infty} c_m (z-a)^m \frac{dz}{(z-a)^{n+1}} = 2\pi i c_n.$$

Thus, $|c_n| \le \frac{1}{2\pi} \int_{\gamma_\rho} \frac{|f(z)|}{\rho^{n+1}} |dz| = \frac{1}{2\pi} \frac{M}{\rho^{n+1}} 2\pi\rho = \frac{M}{\rho^n}.$

Theorem 2.4 (Liouville) *A function that is holomorphic in the whole plane and bounded is a constant.*

Proof Let $f(z) = \sum_{n=0}^{\infty} c_n z^n$ and $|f(z)| \le M$. Then, by Cauchy's inequality, $|c_n| \le \frac{M}{\rho^n}$ for all $\rho > 0$.

2.2 Isolated Singularities

Definition 2.2 A point $a \in \mathbb{C}$ is called an *isolated singularity* of a function $f(z)$ if f is holomorphic in a punctured neighborhood $\{z \in | 0 < |z-a| < r\}$ of a. An isolated singularity a is called *removable* if $\lim_{z\to a} f(z) = A \in \mathbb{C}$; a *pole* if $\lim_{z\to a} f(z) = \infty$; *essential* in the remaining cases.

Theorem 2.5 *Let a be an isolated singularity of a function $f(z) = \sum_{n=-\infty}^{\infty} c_n (z - a)^n$. Then*

(1) the following conditions are equivalent:

- *(a) a is a removable singularity;*
- *(b) $|f(z)| \le M$ in some neighborhood of a;*
- *(c) $c_n = 0$ for all $n < 0$;*

thus, a meromorphic function with removable singularities becomes holomorphic at all points if we redefine it appropriately at these singularities;

(2) the following conditions are equivalent:

- *(a) a is a pole;*
- *(b) there exists $N < 0$ such that $c_N \ne 0$ and $c_n = 0$ for all $n < N$.*

Proof

1. Obviously, (a) implies (b) and (c) implies (a). Let us prove that (b) implies (c). Let $|f(z)| \le M$ in some neighborhood of a. Then, by Cauchy's inequality, $|c_n| \le \frac{M}{\rho^n}$ for $0 < \rho < 1$. Therefore, $c_n = 0$ for all $n < 0$, i.e., (b) implies (c).

2. Let a be a pole. Then $|f(z)| \neq 0$ in some punctured neighborhood of a, and hence the function $\varphi(z) = \frac{1}{f(z)}$ is holomorphic in this neighborhood. By the first part of the theorem already proved, this implies that

$$\varphi(z) = (z-a)^{-N} \cdot \sum_{n=0}^{\infty} c_n(z-a)^n \quad \text{where } c_0 \neq 0 \text{ and } N < 0.$$

Thus, $f(z) = \frac{1}{\varphi(z)} = (z-a)^N \cdot \sum_{n=0}^{\infty} b_n(z-a)^n$ where $b_0 \neq 0$. The converse is obvious.

Definition 2.3 Let $f(z) = (z-a)^N \cdot \sum_{n=0}^{\infty} c_n(z-a)^n$ with $c_0 \neq 0$. If $N > 0$, then N is called the *order of zero* of the function $f(z)$ at the point a; if $N < 0$, then $-N$ is called the *order of pole* of the function $f(z)$ at the point a.

Exercise 2.1 Show that a function f has a pole of order N at a point a if and only if the function f^{-1} has a zero of order N at a.

Theorem 2.6 (Sokhotski–Casorati–Weierstrass Theorem) *Let a be an essential singularity of a function f and $A \in \mathbb{C} \cup \infty$. Then there exists a sequence $a_n \to a$ such that $f(a_n) \to A$.*

Proof Let $A = \infty$. Then the conclusion of the theorem follows from the fact that the function f is unbounded in any neighborhood of a. Let $A \in \mathbb{C}$. Then either (1) there exists a sequence $a_n \to a$ such that $f(a_n) = A$, or (2) there exists a punctured neighborhood of a in which $f(z) \neq A$. In the latter case, the function $\varphi(z) = \frac{1}{f(z)-A}$ is holomorphic in this neighborhood and a is an essential singularity of $\varphi(z)$. Therefore, as we have already proved, there exists a sequence $a_n \to a$ such that $\varphi(a_n) \to \infty$. But then we have $\lim_{n\to\infty} f(a_n) = \lim_{n\to\infty} \left(A + \frac{1}{\varphi(a_n)} \right) = A$.

Definition 2.4 The point ∞ is said to be an *isolated singularity* of a function $f(z)$ if f is holomorphic in a punctured neighborhood $\{R < |z| < \infty\}$ of ∞. In this case, the same classification holds: the singularity is either *removable*, or a *pole*, or an *essential singularity*.

Exercise 2.2 Show that ∞ is an isolated singularity of a function $f(z)$ if and only if 0 is an isolated singularity of the function $g(z) = f(z^{-1})$.

Definition 2.5 A function holomorphic in the whole plane $\mathbb{C}$ is said to be *entire*. A function f on a domain $D \subset \bar{\mathbb{C}} = \mathbb{C} \cup \infty$ whose only singular points are removable singularities and poles is said to be *meromorphic on D*.

Exercise 2.3 Show that a function $f(z)$ is meromorphic on the whole Riemann sphere $\bar{\mathbb{C}}$ if and only if it is rational, i.e.,

$$f(z) = \frac{a_0 z^n + \dots + a_n}{b_0 z^m + \dots + b_m}.$$

This important statement illustrates the fundamental relation between analytic properties of a function and its algebraic properties.

2.3 Residues and Principal Value Integrals

In what follows, unless otherwise stated, all contours inside a subset of $\mathbb{C}$ are oriented counterclockwise.

Definition 2.6 Let $f(z) = \sum_{n=-\infty}^{\infty} c_n (z - z_0)^n$ be a function defined on a domain U and $\gamma = \{z \in \mathbb{C} \mid |z - z_0| = r\} \subset U$. Then the value $\mathrm{Res}_{z_0} f = c_{-1} = \frac{1}{2\pi i} \int_\gamma f(z)\,dz$ is called the *residue* of f at z_0.

Theorem 2.7 *Assume that a function $f(z)$ is holomorphic on a domain D except for isolated singularities, $G \subset D$ is a compact subset, and the boundary ∂G contains no singularities. Then*

$$\frac{1}{2\pi i} \int_{\partial G} f(z)\,dz = \sum_j \mathrm{Res}_{z_j} f$$

where the sum is taken over all singularities lying inside G.

Proof Let $\gamma_j \subset G$ be pairwise disjoint closed contours around the points z_j (see Fig. 2.1). Then, by Theorem 1.6 and Example 1.1, we have

$$\int_{\partial G} f(z)\,dz = \sum_j \int_{\gamma_j} f(z)\,dz = 2\pi i \sum_j \mathrm{Res}_{z_j} f.$$

Definition 2.7 Let $f(z) = \sum_{n=-\infty}^{\infty} c_n z^n$ be a function holomorphic on the domain $U = \{z \in \mathbb{C} \mid |z| > R\}$, and let the contour $\gamma = \{z \in \mathbb{C} \mid |z| = r\} \subset U$ be oriented

Fig. 2.1

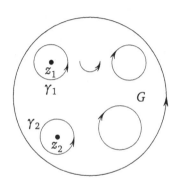

counterclockwise. Then the value

$$\text{Res}_\infty f = -c_{-1} = -\frac{1}{2\pi i} \int\limits_\gamma f(z)\,dz$$

is called the *residue* of the function $f(z)$ at ∞.

Theorem 2.8 *Assume that a function $f(z)$ is holomorphic on the sphere $\bar{\mathbb{C}}$ except for finitely many points $z_1, \ldots, z_n \in \mathbb{C}$. Then*

$$\sum_{i=1}^n \text{Res}_{z_i} f = 0.$$

As you have already noticed, all definitions and theorems for functions defined on domains $D \subset \mathbb{C}$ can be naturally extended to domains containing ∞. From this point of view, the minus sign in the previous definition is explained by the fact that the contour under consideration goes clockwise around ∞.

Definition 2.8 Let $f(z) = \sum\limits_{n=-\infty}^{\infty} c_n(z - z_0)^n$ be a function defined on a domain $G \setminus z_0$ and z_0 be an interior point of a compact curve Γ. Set $G_\varepsilon = \{z \in G \mid |z - z_0| \leq \varepsilon\}$ and $\Gamma_\varepsilon = \Gamma \setminus G_\varepsilon$. Then the limit

$$\text{v. p.} \int\limits_\Gamma f(z)\,dz = \lim_{\varepsilon \to 0} \int\limits_{\Gamma_\varepsilon} f(z)\,dz$$

is called the *principal value* of the integral of f over Γ.

Theorem 2.9 *Let Γ be a smooth curve and $z_0 \in \Gamma$, and let $\mu(z)$ be a function satisfying the Lipschitz condition $|\mu(z) - \mu(z_0)| \leq \text{const}\,|z - z_0|$ and holomorphic on a punctured neighborhood of z_0. Then*

$$\text{v. p.} \int\limits_\Gamma \frac{\mu(z)}{z - z_0}\,dz = \pi i\,\mu(z_0) + \int\limits_\Gamma \frac{\mu(z) - \mu(z_0)}{z - z_0}\,dz.$$

Proof We may assume that the curve Γ is closed. Consider a contour γ around the point z_0 lying in the punctured neighborhood of z_0 in which the function $\mu(z)$ is holomorphic. Set

$$\gamma_\varepsilon = \{z \in \mathbb{C} \mid |z - z_0| = \varepsilon\} = \gamma' \cup \gamma''$$

Fig. 2.2

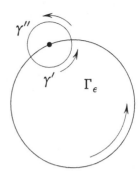

where $\gamma' \cap \gamma'' \subset \Gamma$ (see Fig. 2.2). Then

$$\int_{\Gamma_\varepsilon} \frac{\mu(z)}{z - z_0}\, dz = \int_{\Gamma_\varepsilon} \frac{\mu(z) - \mu(z_0)}{z - z_0}\, dz + \int_{\Gamma_\varepsilon} \frac{\mu(z_0)}{z - z_0}\, dz$$

$$= \int_{\Gamma_\varepsilon} \frac{\mu(z) - \mu(z_0)}{z - z_0}\, dz + \int_{\gamma'} \frac{\mu(z_0)}{z - z_0}\, dz.$$

Hence

$$\lim_{\varepsilon \to 0} \int_{\Gamma_\varepsilon} \frac{\mu(z)}{z - z_0} = \int_{\Gamma} \frac{\mu(z) - \mu(z_0)}{z - z_0}\, dz + \lim_{\varepsilon \to 0} \frac{1}{2} \int_{\gamma_\varepsilon} \frac{\mu(z_0)}{z - z_0}\, dz$$

$$= \int_{\Gamma} \frac{\mu(z) - \mu(z_0)}{z - z_0}\, dz + \pi i \mu(z_0).$$

2.4   The Argument Principle

Definition 2.9 If a function f has a zero (or a pole) of order n at a point z_0, we say that f has n *zeros* (respectively, n *poles*) at z_0.

Theorem 2.10 *Let f be a function meromorphic on a domain D and $\Gamma \subset D$ be a closed contour bounding a set G. Let N and P be the number of zeros and poles of f inside G, and assume that the boundary $\partial G = \Gamma$ contains no zeros or poles of f. Then*

$$N - P = \frac{1}{2\pi i} \int_{\Gamma} \frac{f'(z)}{f(z)}\, dz.$$

Proof Let z_0 be a zero of order n or a pole of order $-n$. Then

$$f(z) = (z - z_0)^n \varphi(z) \quad \text{where} \quad \varphi(z_0) \neq 0,$$
$$f'(z) = (z - z_0)^{n-1} \big((z - z_0) \varphi'(z) + n \varphi(z) \big),$$

and

$$\frac{f'}{f} = \frac{1}{z - z_0} \left(n + (z - z_0) \frac{\varphi'(z)}{\varphi(z)} \right).$$

Thus, $\frac{1}{2\pi i} \int_{\gamma_{z_0}} \frac{f'}{f} \, dz = n$ where $\gamma_{z_0} \subset G$ is a contour separating the point z_0 from the other zeros and poles of f. By Cauchy's theorem, the integral $\frac{1}{2\pi i} \int_{\Gamma} \frac{f'(z)}{f(z)} \, dz$ is equal to the sum of all integrals of the form $\frac{1}{2\pi i} \int_{\gamma_{z_j}} \frac{f'}{f} \, dz$ corresponding to the zeros and poles z_j of f. Therefore, $\frac{1}{2\pi i} \int_{\Gamma} \frac{f'(z)}{f(z)} \, dz = N - P$.

Recall that, given a complex number $u = re^{i\varphi}$, the number $\varphi \in [0, 2\pi)$ is called the *argument* of u and denoted by $\arg u$. Let $f(u)$ be a function defined on a contour Γ such that $f|_{\Gamma} \neq 0$. If u travels counterclockwise around the contour Γ once, then the number $e^{i \arg f(u)}$ travels around the contour $S = \{z \in \mathbb{C} \mid |z| = 1\}$ an integer number of times, which is denoted by $\frac{1}{2\pi} \Delta_{\Gamma} \arg f$.

Theorem 2.11 (Argument Principle)   *Under the assumptions of Theorem 2.10,*

$$N - P = \frac{1}{2\pi} \Delta_{\Gamma} \arg f(z).$$

Proof Consider a deformation of Γ into small contours γ_i around the zeros and poles of $f(z)$ and line segments connecting them with Γ (see Fig. 1.2 at page 9). Each segment is traversed twice in opposite directions and does not contribute to $\frac{1}{2\pi} \Delta_{\Gamma} \arg f(z)$. Therefore, $\frac{1}{2\pi} \Delta_{\Gamma} \arg f(z)$ is equal to the sum of the values $\frac{1}{2\pi} \Delta_{\gamma_i} \arg f(z)$ over all contours γ_i. If γ is a small contour around a point z_0 where $f(z) = (z - z_0)^n \varphi(z)$ and $\varphi(z_0) \neq 0$, then

$$\frac{1}{2\pi} \Delta_{\gamma} \arg f(z) = \frac{1}{2\pi} \Delta_{\gamma} \arg(z - z_0)^n = n.$$

Theorem 2.12 (Rouché)   *Let f and g be functions holomorphic on a domain D, and let $\Gamma \subset D$ be a closed contour that bounds a set G and contains no zeros of f. If $|f(z)| > |g(z)|$ on Γ, then the functions f and $f + g$ have the same number of zeros inside G.*

Proof Set $F_\lambda = f + \lambda g$. Then $F_\lambda|_{\partial G} \neq 0$ for $0 \leq \lambda \leq 1$. It follows that the function $\psi(\lambda) = \frac{1}{2\pi} \Delta_\Gamma \arg F_\lambda(z)$ exists, is continuous, and, therefore, constant. In particular,

$$\frac{1}{2\pi} \Delta_\Gamma \arg f(z) = \frac{1}{2\pi} \Delta_\Gamma \arg F_0(z) = \frac{1}{2\pi} \Delta_\Gamma \arg F_1(z) = \frac{1}{2\pi} \Delta_\Gamma \arg(f(z) + g(z)).$$

The argument principle completes the proof.

Corollary 2.1 (Fundamental Theorem of Algebra) *A polynomial of degree n has exactly n roots in $\mathbb{C}$.*

Proof An arbitrary polynomial has the form

$$P_n = a_n z^n + \ldots + a_0 = f(z) + g(z),$$

where $f(z) = a_n z^n$ and $g(z) = a_{n-1} z^{n-1} + \ldots + a_0$. Now, apply Rouché's theorem to the pair of functions f, g and the contour $\Gamma_R = \{z \in \mathbb{C} \mid |z| = R\}$ for sufficiently large R.

2.5 Topological Properties of Meromorphic Functions

Lemma 2.1 *Let $f(z) = w_0 + (z - z_0)^n \varphi(z)$ where $n \geq 1$, the function φ is holomorphic in a neighborhood of z_0, and $\varphi(z_0) \neq 0$. Then there exist domains U and W such that $z_0 \in U$, $w_0 \in W \subset f(U)$, and for every point $w \in W \setminus w_0$ the function $f|_U$ takes the value w at exactly n different points.*

Proof Choose r such that $\varphi(z)$ does not vanish on the set $D = \{z \in \mathbb{C} \mid |z - z_0| \leq r\}$ and the derivative f' has no zeros in $D \setminus z_0$. Set $U = D \setminus \partial D$, $\mu = \min_{z \in \partial D} |f(z) - w_0| > 0$, and $W = \{w \in \mathbb{C} \mid |w - w_0| < \mu\}$.

For an arbitrary point $w \in W$, consider the function

$$F(z) = f(z) - w = (f(z) - w_0) + (w_0 - w).$$

On the contour ∂D, its parts $f(z) - w_0$ and $w_0 - w$ satisfy the condition

$$|f(z) - w_0| \geq \mu \geq |w_0 - w|.$$

By Rouché's theorem, it follows that the function $F(z) = f(z) - w$ has the same number of zeros in U as the function $f(z) - w_0$, i.e., n zeros. For $w \neq w_0$, all these zeros are distinct, since the derivative f' does not vanish on $U \setminus z_0$.

Theorem 2.13 (Open Mapping Theorem) *If f is a holomorphic function on a domain D and $f \neq$ const, then $f(D)$ is a domain too.*

Proof By Lemma 2.1, for every point $z_0 \in D$ there exists a neighborhood W of the point $w_0 = f(z_0)$ such that $W \subset f(D)$.

Theorem 2.14 (Maximum Modulus Principle) *If a nonconstant function f is holomorphic on a domain D and continuous on the closure $\bar{D} \subset \mathbb{C} \cup \infty$, then*
$$\max_{z \in \bar{D}} |f(z)| = \max_{z \in \partial \bar{D}} |f(z)|.$$

Proof Assume that the function $|f|$ attains the maximum value at a point $z_0 \in D$ and $w_0 = f(z_0)$. Then, by Theorem 2.13, we have $W = \{z \in \mathbb{C} \mid |w - w_0| < r\} \subset f(D)$ for some r. The set W contains points w such that $|w| > |w_0|$. Hence, there exists a point $z \in D$ such that $w = f(z) \in W$ and $|f(z)| = |w| > |w_0|$.

Theorem 2.15 (Schwarz Lemma) *Let $f(z)$ be a holomorphic function on the domain $U = \{z \in \mathbb{C} \mid |z| < 1\}$ such that $f(0) = 0$ and $|f(z)| \leq 1$. Then $|f(z)| \leq |z|$ for all points $z \in U$. If, moreover, $|f(z_0)| = |z_0|$ for some $z_0 \neq 0$, then $f(z) = \alpha z$ where $|\alpha| = 1$.*

Proof The function $\varphi(z) = \frac{f(z)}{z}$ is holomorphic on every disk $U_r = \{z \in \mathbb{C} \mid |z| \leq r\}$ with $r < 1$. By Theorem 2.14, we have $\max_{z \in U_r} |\varphi(z)| \leq \max_{z \in \partial U_r} \left| \frac{f(z)}{z} \right| \leq \frac{1}{r}$. Thus, $|\varphi(z)| \leq 1$, i.e., $|f(z)| \leq |z|$. If $|f(z_0)| = |z_0|$ for $z_0 \in U$, then $z_0 \in U_r \setminus \partial U_r$ and

$$|\varphi(z_0)| = 1 = \max_{z \in D_r} \varphi(z).$$

By Theorem 2.14, it follows that $\varphi(z) = \alpha = \text{const}$ where $|\alpha| = 1$. Thus, $f(z) = \alpha z$.

Chapter 3
Riemann Mapping Theorem

3.1 Continuous Functionals on Compact Families of Functions

Definition 3.1 A family $\mathfrak{F}$ of functions is said to be *uniformly bounded inside a domain D* if for every compact set $K \subset D$ there exists a constant $M = M(K)$ such that $|f(z)| \leq M$ for all $f \in \mathfrak{F}, z \in K$.

Exercise 3.1 Show that if a family $\mathfrak{F}$ of holomorphic functions is uniformly bounded inside a domain D, then the family $\{f'\}$ is also uniformly bounded inside D. (*Hint.* Use Cauchy's formula.)

Definition 3.2 A family of functions $\mathfrak{F}$ is said to be *equicontinuous inside a domain D* if for every $\varepsilon > 0$ and every compact set $K \subset D$ there exists $\delta = \delta(\varepsilon, K)$ such that $|f(z_1) - f(z_2)| < \varepsilon$ for all $f \in \mathfrak{F}$ and for $z_1, z_2 \in K$ such that $|z_1 - z_2| < \delta$.

Exercise 3.2 Show that if a family $\mathfrak{F}$ of functions holomorphic on a domain D is uniformly bounded inside D, then it is equicontinuous inside D. (*Hint.* Use Exercise 3.1.)

Definition 3.3 A sequence of functions on a domain D is said to be *fundamental* if it converges uniformly on every compact set $K \subset D$.

Exercise 3.3 Using Weierstrass' theorem 1.15, show that the limit of a fundamental sequence of functions holomorphic on D is also holomorphic on D.

Theorem 3.1 (Montel) *Let $\mathfrak{F}$ be a family of holomorphic functions uniformly bounded inside a domain D. Then every sequence $\{f_n\}$ in $\mathfrak{F}$ contains a fundamental subsequence.*

Proof Let $Q = \{\tilde{z}_1, \tilde{z}_2, \ldots\} \subset D$ be the subset of all points in D with rational real and imaginary parts. Choose a subsequence $\{f_n^1\}$ of $\{f_n\}$ such that $\{f_n^1(\tilde{z}_1)\}$

© Springer Nature Switzerland AG 2019
S. M. Natanzon, *Complex Analysis, Riemann Surfaces and Integrable Systems*,
Moscow Lectures 3, https://doi.org/10.1007/978-3-030-34640-9_3

converges. Then choose a subsequence $\{f_n^2\}$ of $\{f_n^1\}$ such that $\{f_n^2(\tilde{z}_2)\}$ converges, etc. Set $h_n = f_n^n$. Then the sequence $\{h_n(\tilde{z}_p)\}$ converges for every p. We will prove that the sequence $\{h_n\}$ is fundamental. Let $K \subset D$ be a compact set. Then, by Exercise 3.2, the set K can be covered by squares so that if z' and z'' belong to the same square, then $|f(z') - f(z'')| < \frac{\varepsilon}{3}$. Since K is compact, we may assume that there are finitely many of these squares. In each of them choose a point from the set Q, obtaining points $z_1, \ldots, z_p$. Since the sequences $\{h_n(z_i)\}$ converge for every i, by the Cauchy convergence test there exists N such that $|h_m(z_i) - h_n(z_i)| < \frac{\varepsilon}{3}$ for $n, m > N$ and all i. Thus, if z_k lies in the same square as z, then

$$|h_m(z) - h_n(z)| \le |h_m(z) - h_m(z_k)| + |h_m(z_k) - h_n(z_k)| + |h_n(z_k) - h_n(z)| < \varepsilon.$$

Therefore, by the Cauchy convergence test, the sequence of functions $\{h_n\}$ uniformly converges on K.

Definition 3.4 A family $\mathfrak{F}$ of functions defined on a domain D is said to be *compact* if any sequence of functions $\{f_n\}$ in $\mathfrak{F}$ contains a fundamental subsequence converging to a function from $\mathfrak{F}$.

Definition 3.5 A map $J: \mathfrak{F} \to \mathbb{C}$ defined on a family of functions $\mathfrak{F}$ is called a *functional*. A functional is said to be *continuous* if for every fundamental sequence $\{f_n\}$ in $\mathfrak{F}$ converging to a function $f \in \mathfrak{F}$, we have $\lim_{n\to\infty} J(f_n) = J(f)$.

Exercise 3.4 Let $\mathfrak{F}$ be a family of functions holomorphic on a domain $D \ni a$. Let $J(f) \overset{\text{def}}{=} f^{(p)}(a)$, where $f^{(p)}(a)$ is the pth derivative of f at a. Show that J is a continuous functional.

Exercise 3.5 Show that a continuous functional defined on a compact family of functions is bounded.

Theorem 3.2 *Let J be a continuous functional on a compact family $\mathfrak{F}$ of functions on D. Then there exists a function $f_0 \in \mathfrak{F}$ such that $|J(f_0)| \ge |J(f)|$ for all functions $f \in \mathfrak{F}$.*

Proof Let $A = \sup_{f \in \mathfrak{F}} |J(f)|$. Then there exists a sequence $\{f_n\}$ in $\mathfrak{F}$ such that

$$\lim_{n\to\infty} |J(f_n)| = A.$$

Since $\mathfrak{F}$ is compact, there exists a fundamental subsequence $\{h_m\}$ in $\{f_n\}$ that converges to a function $f_0 \in \mathfrak{F}$. Since the functional J is continuous, we obtain

$$A = \lim_{n\to\infty} |J(f_n)| = \lim_{m\to\infty} |J(h_m)| = |J(f_0)|.$$

3.2 Hurwitz' Theorem and Univalent Functions

Theorem 3.3 (Hurwitz) *Let $\{f_n\}$ be a fundamental sequence of holomorphic functions on a domain D, and let $f = \lim_{n\to\infty} f_n \neq$ const and $f(z_0) = 0$. Then for every $r > 0$ there exists N such that for every $n > N$ the function f_n has a zero in the domain $\{z \in D \mid |z - z_0| < r\}$.*

Proof Using Exercise 3.3 and Theorem 1.14, we see that the function $f(z)$ can be written as $f(z) = (z - z_0)^p(a + \varphi(z))$ where $a \neq 0$, the function $\varphi(z)$ is holomorphic, and $\varphi(z_0) = 0$. Hence, there exists $\rho > 0$ such that $|f(z)| > 0$ on the set $Q = \{0 < |z - z_0| \leq \rho\}$. Put $\mu = \min_{\partial Q} |f(z)| > 0$. Since the sequence $\{f_n\}$ converges uniformly on the boundary ∂Q, there exists N such that for any $n > N$ and $z \in \partial Q$ we have $|f_n(z) - f(z)| < \mu$. Therefore, by Rouché's theorem (2.12), the function $f_n = f + (f_n - f)$ has a zero in the domain $Q \setminus \partial Q$.

Definition 3.6 A function f is said to be *univalent* if it establishes a one-to-one correspondence, i.e., $f(z_1) \neq f(z_2)$ for $z_1 \neq z_2$.

Theorem 3.4 *A holomorphic function f is univalent in a neighborhood of a point z_0 if and only if $f'(z_0) \neq 0$.*

Proof By Lemma 2.1, a function f is invertible in a neighborhood of a point z_0 if and only if $f(z) = w_0 + (z - z_0)\varphi(z)$ where $\varphi(z_0) \neq 0$, which, in turn, is equivalent to the condition that $f'(z_0) \neq 0$.

Exercise 3.6 Show that the function g inverse to a univalent function f is also univalent and $g'(w_0) = \frac{1}{f'(z_0)}$.

Theorem 3.5 *Let $\{f_n\}$ be a fundamental sequence of univalent functions on a domain D that converges to a nonconstant function f. Then f is univalent.*

Proof By Weierstrass' theorem, the function f is holomorphic. Assume that $z_1 \neq z_2$ and $f(z_1) = f(z_2)$. Let $Q = \{z \in D \mid |z - z_1| < |z_2 - z_1|\}$. The sequence of functions $h_n(z) = f_n(z) - f_n(z_2)$ converges to the function $h(z) = f(z) - f(z_2)$, where $h(z_1) = 0$. By Theorem 3.3, there exist N and $z_0 \in Q$ such that $h_N(z_0) = 0$. Therefore, $f_N(z_0) = f_N(z_2)$, which is a contradiction, since f_N is univalent.

Theorem 3.6 *Let S be the family of all holomorphic univalent functions on a domain D satisfying the condition $|f| \leq 1$. Assume that S is nonempty. Then there exist a function $f_0 \in S$ and a point $a \in D$ such that $0 < |f_0(a)|$ and $|f'(a)| \leq |f_0'(a)|$ for all $f \in S$.*

Proof By assumption, there exist a function $f_1 \in S$ and a point $a \in D$ such that $|f_1'(a)| > 0$. Let $S_1 = \{f \in S \mid |f'(a)| \geq |f_1'(a)|\}$. According to Montel's theorem 3.1, every sequence $\{f_n\}$ in S_1 contains a fundamental subsequence. By Weierstrass' theorem, its limit g is a holomorphic function for which $|g'(a)| \geq |f_1'(a)| > 0$, i.e., $g \neq$ const. By Theorem 3.5, it follows that g is a univalent function, i.e., $g \in S_1$. Thus, S_1 is a compact family of functions. Define a functional y on S_1 by the formula $y(a) = |f'(a)|$. By Exercise 3.4, this functional is continuous; by Theorem 3.2, it attains the maximum value at a function $f_0 \in S_1$. Therefore, $|f_0'(a)| \geq |f'(a)|$ for all $f \in S$.

3.3   Analytic Continuation

Definition 3.7 A *canonical element* is a pair (U_a, f_a) where U_a is a disk centered at a point a and $f_a(z) = \sum\limits_{i=0}^{\infty} c_i(z - a)^i$ is a convergent series in U_a. Canonical elements (U_a, f_a) and $(\tilde{U}_a, \tilde{f}_a)$ are said to be *equivalent* if $f_a|_{U_a \cap \tilde{U}_a} = \tilde{f}_a|_{U_a \cap \tilde{U}_a}$.

Definition 3.8 Let γ be a non-self-intersecting path in $\mathbb{C}$ connecting points $a \neq b$. A canonical element (U_b, f_b) is called an *analytic continuation of a canonical element* (U_a, f_a) along the path γ if there exist canonical elements $(\tilde{U}_a, \tilde{f}_a)$ and $(\tilde{U}_b, \tilde{f}_b)$ equivalent to (U_a, f_a) and (U_b, f_b), respectively, a domain $D \supset \gamma \cup \tilde{U}_a \cup \tilde{U}_b$, and a holomorphic function $f \colon D \to \mathbb{C}$ such that $f|_{\tilde{U}_a} = \tilde{f}_a$ and $f|_{\tilde{U}_b} = \tilde{f}_b$.

Definition 3.9 In what follows, by a *path* we always mean a path that can be divided into finitely many non-self-intersecting segments. The analytic continuation along such a path is defined as the composition of the analytic continuations along its non-self-intersecting parts.

Exercise 3.7 Show that the analytic continuation along a path does not depend on the partition of the path into non-self-intersecting segments.

Remark 3.1 The analytic continuation along a path γ can be constructed as follows. Cover γ by disks U_{a_i}, $1 \leq i \leq k$ (here $\gamma \subset \bigcup\limits_{i=1}^{k} U_{a_i}$, $U_{a_1} = U_a$, and $U_{a_k} = U_b$), and successively "reexpand" the function f, passing from disk to disk and constructing canonical elements (U_{a_i}, f_{a_i}) in such a way that

$$f_{a_i}|_{U_{a_i} \cap U_{a_{i+1}}} = f_{a_{i+1}}|_{U_{a_i} \cap U_{a_{i+1}}}.$$

Theorem 3.7 *Let γ_0 and γ_1 be homotopic paths with the same endpoints a and b, and let γ_t ($t \in [0, 1]$) be a homotopy between them. Let (U_a, f_a) be a canonical element that can be analytically continued along each path γ_t. Then the analytic continuations of the canonical element (U_a, f_a) along the paths γ_0 and γ_1 are equivalent.*

Proof Let $(U_{a_i^t}, f_{a_i^t})$ be the canonical elements corresponding to the path γ_t and the points $a_1, \ldots, a_{k_t}$ from Remark 3.1. Consider the set T of points $t \in [0, 1]$ such that the analytic continuations along the paths γ_0 and γ_t are equivalent. For every domain $D \supset \gamma_t$ there exists $\delta > 0$ such that $D \supset \gamma_{t'}$ for $|t - t'| < \delta$. Thus, the set T is open. For obvious reasons, T is closed. Hence, $T = [0, 1]$.

Exercise 3.8 Show that if the assumptions of Theorem 3.7 on the existence of analytic continuations along each path γ_t are violated, then the analytic continuations along the paths γ_0 and γ_1 can be nonequivalent.

3.4   Riemann Mapping Theorem

Exercise 3.9 Show that if $|a| < 1$, $|b| < 1$, then $\left| \frac{a-b}{1-\bar{a}b} \right| < 1$.

Definition 3.10 Domains D_1, $D_2 \subset \bar{\mathbb{C}}$ are said to be *biholomorphically equivalent* if there is a one-to-one holomorphic map $\varphi \colon D_1 \to D_2$ between them. In this case, the inverse map $\varphi^{-1} \colon D_2 \to D_1$ is also holomorphic (Exercise 3.6). For this reason, such a map φ is said to be *biholomorphic*.

Example 3.1

- The formula $h(z) = \frac{1}{z} + c$ defines a biholomorphic map from $\mathbb{C}$ to $\bar{\mathbb{C}} \setminus c$. Therefore, all Riemann spheres with one point removed are biholomorphically equivalent to $\mathbb{C}$.
- A biholomorphic map is a homeomorphism. Therefore, the Riemann sphere $\bar{\mathbb{C}} = \mathbb{C} \cup \infty$ is not biholomorphically equivalent to the complex plane $\mathbb{C}$ and to the unit disk $\Lambda = \{z \in \mathbb{C} \mid |z| < 1\}$.
- The complex plane $\mathbb{C}$ and the unit disk Λ are homeomorphic, but still not biholomorphically equivalent. This follows from the following remark.

    The map $f(z) \mapsto f^*(z) = f(\varphi(z))$ establishes a one-to-one correspondence between the sets of holomorphic functions on D_1 and D_2. It takes bounded functions to bounded functions and constant functions to constant functions. The function $f(z) = z$ is holomorphic, bounded, and nonconstant on Λ. At the same time, by Liouville's theorem, all bounded holomorphic functions on $\mathbb{C}$ are constant.

Theorem 3.8 (Riemann) *Every connected, simply connected domain on the Riemann sphere is biholomorphically equivalent to either the Riemann sphere $\bar{\mathbb{C}}$ itself, the complex plane $\mathbb{C}$, or the unit disk Λ.*

Proof Assume that a connected, simply connected domain $D \subset \bar{\mathbb{C}}$ is biholomorphically equivalent neither to $\bar{\mathbb{C}}$ nor to $\mathbb{C}$. Then the complement $\bar{\mathbb{C}} \setminus D$ contains distinct points $\alpha \neq \beta$. At a point $a \in D$, the function $f = \sqrt{\frac{z-\alpha}{z-\beta}}$ takes two values and generates two canonical elements (U_a^1, f_a^1), (U_a^2, f_a^2), where $f_a^1 = -f_a^2$ on $U_a^1 \cap U_a^2$. Connect an arbitrary point $b \in D$ with a by a path γ in D. Let (U_b^i, f_b^i) be the analytic continuation of the element (U_a^i, f_a^i) along γ. By Theorem 3.7, the canonical element (U_b^i, f_b^i) does not depend on γ. Hence, there exist analytic functions $f^i \colon D \to \mathbb{C}$ such that $f_b^i = f^i|_{U_b}$ for all $b \in D$, and, moreover, $f^2 = -f^1$. Put $D_i = f^i(D)$. If $f^i(z_1) = \pm f^i(z_2)$, then $\frac{z_1-\alpha}{z_1-\beta} = \frac{z_2-\alpha}{z_2-\beta}$, whence $z_1 = z_2$.

Thus, the functions f^i are univalent and $D_1 \cap D_2 = \varnothing$. By the open mapping Theorem 2.13, the domain D_2 contains a disk $W = \{w \in \mathbb{C} \mid |w - w_0| < \rho\}$; we have $|f^1(z) - w_0| \geq \rho$, since $W \cap D_1 = \varnothing$. Put $\tilde{f}(z) = \frac{\rho}{f^1(z) - w_0}$. The function $\tilde{f} \neq \text{const}$ is holomorphic, univalent, and $|\tilde{f}(z)| \leq 1$.

Consider the set S of all univalent functions $g \colon D \to \Lambda$. It contains the nonconstant function $\tilde{f}$. By Theorem 3.6, there exist a point $a \in D$ and a function $f_0 \in S$ such that $|g'(a)| \leq |f_0'(a)| > 0$ for all functions $g \in S$.

Let us prove that $f_0(D) = \Lambda$. Set $h(z) = \frac{f_0(z) - f_0(a)}{1 - \overline{f_0(a)} f_0(z)}$. Then the function $h(z)$ is univalent, and, by Exercise 3.9, we have $|h(z)| \leq 1$. Therefore, $h \in S$, whence $|f_0'(a)| \geq |h'(a)| = \frac{1}{1 - |f_0(a)|^2} |f_0'(a)|$ and $f_0(a) = 0$.

Let us prove that every point $b \in \Lambda \setminus 0$ belongs to $f_0(D)$. Assume that $b \notin f_0(D)$. Consider the function $\psi(z) = \sqrt{\frac{f_0(z) - b}{1 - \overline{b} f_0(z)}}$. Taking the analytic continuation of the canonical element (U_a, ψ_a^1) of ψ, we construct a holomorphic function ψ^1 on D. Consider the function $\tilde{h}(z) = \frac{\psi^1(z) - \psi^1(a)}{1 - \overline{\psi^1(a)} \psi^1(z)}$. It is univalent and, by Exercise 3.9, we have $|\tilde{h}(z)| \leq 1$, i.e., $\tilde{h} \in S$. But then $|f_0'(a)| \geq |\tilde{h}'(a)| = \frac{1 + |b|}{2\sqrt{|b|}} |f_0'(a)| > |f_0'(a)|$. The obtained contradiction shows that $b \in f_0(D)$.

3.5 Automorphisms of Simply Connected Domains

A holomorphic isomorphism of a domain $D \subset \bar{\mathbb{C}}$ to itself is called an *automorphism* of D. The set $\text{Aut}(D)$ of such automorphisms is a group under composition.

Exercise 3.10 Show that the map $f(z) = e^{i\alpha} \frac{z - a}{1 - \overline{a}z}$ lies in the group $\text{Aut}(\Lambda)$ if $|a| < 1$.

Theorem 3.9 *The following relations hold:*

$$\text{Aut}(\bar{\mathbb{C}}) = \left\{ z \mapsto \frac{az + b}{cz + d} \mid a, b, c, d \in \mathbb{C}, \ ad - bc \neq 0 \right\},$$

$$\text{Aut}(\mathbb{C}) = \left\{ z \mapsto az + b \mid a, b \in \mathbb{C}, \ a \neq 0 \right\},$$

$$\text{Aut}(\Lambda) = \left\{ z \mapsto e^{i\alpha} \frac{z - a}{1 - \overline{a}z} \mid |a| < 1, \ \alpha \in \mathbb{R} \right\}.$$

Proof It follows from definitions that $\text{Aut}(\mathbb{C}) = \left\{ f \in \text{Aut}(\bar{\mathbb{C}}) \mid f(\infty) = \infty \right\}$. Let $g \in \text{Aut}(\bar{\mathbb{C}}) \setminus \text{Aut}(\mathbb{C})$. Then $g(a) = \infty$ for $a \in \mathbb{C}$. By Theorems 2.5 and 2.7 on isolated singularities, the function $g(z)$ has the form $g(z) = \frac{A}{z - a} + \varphi(z)$ where φ is a holomorphic function on $\mathbb{C}$. Besides, $\lim_{z \to \infty} \varphi(z) = \lim_{z \to \infty} g(z) = g(\infty) \in \mathbb{C}$, and, by Liouville's theorem 2.4, we obtain $\varphi(z) = \text{const}$. Thus,

$$\text{Aut}(\bar{\mathbb{C}}) \setminus \text{Aut}(\mathbb{C}) = \left\{ z \mapsto \frac{A}{z-a} + B \mid A, B, a \in \mathbb{C}, \ A \neq 0 \right\}.$$

If $f \in \text{Aut}(\mathbb{C})$, then $f(z^{-1}) \in \text{Aut}(\bar{\mathbb{C}}) \setminus \text{Aut}(\mathbb{C})$ and, as we have already proved, $f(z^{-1}) = \frac{A}{z} + B$. It follows that $f(z) = Az + B$ with $A \neq 0$.

Let $f \in \text{Aut}(\Lambda)$ and $f(a) = 0$. Set $\varphi(z) = \frac{z-a}{1-\bar{a}z}$ and $g(z) = f(\varphi^{-1}(z))$. Then $g \in \text{Aut}(\Lambda)$ by Exercise 3.10 and $g(0) = 0$. Applying Schwarz's lemma (Theorem 2.15) to the functions g and g^{-1}, we see that $|g(z)| = |z|$. Hence, by Schwarz's lemma, $g(z) = e^{i\alpha}z$. Thus, for $w = \varphi(z)$ we have $f(w) = g(\varphi(w)) = e^{i\alpha}\frac{w-a}{1-\bar{a}w}$.

Exercise 3.11 Show that

$$\text{Aut}(\{z \in \mathbb{C} \mid \Im z > 0\}) = \left\{ z \mapsto \frac{az+b}{cz+d} \mid a, b, c, d \in \mathbb{R}, \ ad - bc > 0 \right\}.$$

3.6 Carathéodory's Theorem

Consider the problem of extending a biholomorphic map to the boundary. The following important theorem, which we state without proof, is known as Carathéodory's theorem.

Theorem 3.10 (Carathéodory [24, Part 1, Sec. 12]) *Let $D_1, D_2 \subset \mathbb{C}$ be domains bounded by Jordan curves. Then a biholomorphic map $f: D_1 \to D_2$ extends to a homeomorphism $\bar{f}: \bar{D}_1 \to \bar{D}_2$ of their closures.*

Exercise 3.12 Show that if the boundary of a domain D contains an analytic arc γ, then a biholomorphic map from D to the unit disk can be analytically continued through γ.

Chapter 4
Harmonic Functions

4.1 Holomorphic and Harmonic Functions

As before, we identify the real plane $\mathbb{R} \times \mathbb{R} = \{(x, y)\}$ with the complex plane $\mathbb{C} = \{z\}$ setting $z = x + iy$. Recall that an open connected subset $D \subset \mathbb{C} = \mathbb{R}^2$ is called a *domain*.

Definition 4.1 A real function $u(x, y)$ on a domain D with continuous second-order partial derivatives is said to be *harmonic* if it satisfies the Laplace equation $\Delta u = 0$ where

$$\Delta = \frac{\partial^2}{\partial x^2} + \frac{\partial^2}{\partial y^2} = \frac{1}{4} \frac{\partial^2}{\partial z \, \partial \bar{z}}$$

is the Laplace operator.

Harmonic functions arise naturally in a wide class of applications, from hydromechanics to theoretical physics.

Theorem 4.1 *A function u on a simply connected domain D is harmonic if and only if it coincides with the real part of a holomorphic function f, i.e., $u(x, y) = \Re f(x + iy)$.*

Proof Let $f(x + iy) = u(x, y) + iv(x, y)$ be a holomorphic function. Then, by the Cauchy–Riemann equations, we have

$$\frac{\partial u}{\partial x} = \frac{\partial v}{\partial y}, \quad \frac{\partial u}{\partial y} = -\frac{\partial v}{\partial x}.$$

© Springer Nature Switzerland AG 2019

S. M. Natanzon, *Complex Analysis, Riemann Surfaces and Integrable Systems*,
Moscow Lectures 3, https://doi.org/10.1007/978-3-030-34640-9_4

Therefore,

$$\frac{\partial^2 u}{\partial x^2} + \frac{\partial^2 u}{\partial y^2} = \frac{\partial}{\partial x}\frac{\partial u}{\partial x} + \frac{\partial^2 u}{\partial y^2} = \frac{\partial}{\partial x}\frac{\partial v}{\partial y} + \frac{\partial^2 u}{\partial y^2} = \frac{\partial}{\partial y}\frac{\partial v}{\partial x} + \frac{\partial^2 u}{\partial y^2}$$

$$= -\frac{\partial}{\partial y}\frac{\partial u}{\partial y} + \frac{\partial^2 u}{\partial y^2} = 0.$$

Let

$$\frac{\partial^2 u}{\partial x^2} + \frac{\partial^2 u}{\partial y^2} = 0.$$

Then

$$\frac{\partial}{\partial x}\left(\frac{\partial u}{\partial x}\right) = \frac{\partial}{\partial y}\left(-\frac{\partial u}{\partial y}\right) \quad \text{and} \quad \frac{\partial}{\partial x}\left(-\frac{\partial u}{\partial y}\right) = -\frac{\partial}{\partial y}\left(\frac{\partial u}{\partial x}\right),$$

which is equivalent to the Cauchy–Riemann equations for the function

$$g(x + iy) = \frac{\partial u}{\partial x} - i\frac{\partial u}{\partial y}.$$

Thus, $g(z)$ is a holomorphic function. Consider its antiderivative

$$f(x + iy) = \int_{z_0}^{z} g(z)\,dz = \int_{(x_0, y_0)}^{(x,y)} \left(\frac{\partial u}{\partial x} - i\frac{\partial u}{\partial y}\right)(dx + i\,dy).$$

Then

$$\Re f(x + iy) = u(x_0, y_0) + \int_{(x_0, y_0)}^{(x,y)} \left(\frac{\partial u}{\partial x}\,dx + \frac{\partial u}{\partial y}\,dy\right) = u(x, y).$$

Exercise 4.1 Show that a function u on a simply connected domain D is harmonic if and only if it coincides with the imaginary part of a holomorphic function.

A close relation between harmonic and holomorphic functions allows one to easily extend many properties of holomorphic functions to harmonic functions.

Exercise 4.2 Show that on a domain D,

- a harmonic function is infinitely differentiable, and all its partial derivatives are also harmonic;

- a biholomorphic change of domain takes a harmonic function to a harmonic function;
- harmonic functions on a connected set A coincide if they coincide on an open subset of A.

4.2   Integral Formulas

In what follows, we assume that the boundary ∂D of the domain under consideration is an analytic curve oriented so that as ∂D is traversed in the positive direction, the domain D is on the left. By

$$\frac{\partial}{\partial n} = \cos\theta\frac{\partial}{\partial x} + \sin\theta\frac{\partial}{\partial y}$$

we denote the directional derivative in the direction of the outer normal $n = (\cos\theta, \sin\theta)$ to the boundary ∂D.

Exercise 4.3   Using Stokes' theorem and the relation

$$\frac{\partial\varphi}{\partial n}ds = -\frac{\partial\psi}{\partial x}dx + \frac{\partial\psi}{\partial y}dy,$$

prove Green's theorem

$$\oint_{\partial D}\varphi\frac{\partial\psi}{\partial n}ds = \iint_D\left(\frac{\partial\varphi}{\partial x}\frac{\partial\psi}{\partial x} + \frac{\partial\varphi}{\partial y}\frac{\partial\psi}{\partial y}\right)dx\,dy + \iint_D\varphi\left(\frac{\partial^2\psi}{\partial x^2} + \frac{\partial^2\psi}{\partial x^2}\right)dx\,dy$$

for functions φ, ψ twice continuously differentiable on D.

Theorem 4.2   *Let* $u(x + iy) = u(x, y)$, $v(x + iy) = v(x, y)$ *be functions that are harmonic on* D *and twice continuously differentiable on* $\bar{D}$. *Then*

$$\oint_{\partial D}\left(u\frac{\partial v}{\partial n} - v\frac{\partial u}{\partial n}\right)ds = 0,$$

$$u(\xi) = \frac{1}{2\pi\rho}\oint_{|z-\xi|=\rho}u(z)\,ds,$$

and

$$\oint_{z\in\partial D}\left\{u(z)\frac{\partial}{\partial n}\ln|z-\xi| - \frac{\partial u}{\partial n}\ln|z-\xi|\right\}ds = \delta 2\pi u(\xi),$$

where $\delta = 1$ *for* $\xi \in D$ *and* $\delta = 0$ *for* $\xi \in \mathbb{C}\setminus\bar{D}$.

Proof The first formula immediately follows from Exercise 4.3. Consider the function

$$v(z) = \Re(\ln(z - \xi)) = \ln|z - \xi|.$$

For $0 < |z - \xi| < \infty$, it is the real part of a holomorphic function, so v is a harmonic function on $\mathbb{C} \setminus \xi$. Therefore, it follows from the part already proved that

$$\oint_{\partial D} \left\{ u(z) \frac{\partial}{\partial n} \ln|z - \xi| - \frac{\partial u}{\partial n} \ln|z - \xi| \right\} ds = 0$$

for $\xi \in \mathbb{C} \setminus \bar{D}$.

Now let $\xi \in D$. Consider the neighborhood $U_\rho = \{z \in \mathbb{C} \mid |z - \xi| < \rho\} \subset D$ and the domain $D_\rho = D \setminus U_\rho$. Then $\xi \in \mathbb{C} \setminus \bar{D}_\rho$, and it follows from the claim just proved that

$$\oint_{\partial D_\rho} \left\{ u(z) \frac{\partial}{\partial n} \ln|z - \xi| - \frac{\partial u}{\partial n} \ln|z - \xi| \right\} ds = 0.$$

Thus,

$$\oint_{\partial D} \left\{ u(z) \frac{\partial}{\partial n} \ln|z - \xi| - \frac{\partial u}{\partial n} \ln|z - \xi| \right\} ds$$

$$= \oint_{\partial U_\rho} u(z) \frac{\partial}{\partial n} \ln|z - \xi| \, ds - \oint_{\partial U_\rho} \frac{\partial u}{\partial n} \ln|z - \xi| \, ds.$$

On the other hand, $\ln|z - \xi| = \rho$ on ∂U_ρ and, consequently, the second integral vanishes by the first formula of the theorem, which we have already proved. Besides,

$$\frac{\partial}{\partial n} \ln|z - \xi| = \frac{\partial}{\partial r} \ln r \big|_{r=\rho} = \frac{1}{\rho}$$

and, consequently,

$$\oint_{\partial D} \left\{ u(z) \frac{\partial}{\partial n} \ln|z - \xi| - \frac{\partial u}{\partial n} \ln|z - \xi| \right\} ds = \frac{1}{\rho} \oint_{\partial U_\rho} u(z) \, ds.$$

The left-hand side of this equality does not change as $\rho \to 0$, whence

$$\oint_{\partial D} \left\{ u(z) \frac{\partial}{\partial n} \ln|z - \xi| - \frac{\partial u}{\partial n} \ln|z - \xi| \right\} ds = \frac{1}{\rho} \oint_{\partial U_\rho} u(z) \, ds = 2\pi u(\xi).$$

4.3 Green's Function

Definition 4.2 A *Green's function* $G = G_D$ on a domain $D \subset \bar{\mathbb{C}}$ is a function $G(z, \xi) = \frac{1}{2\pi} \ln|z - \xi| + g(z, \xi)$ on $\bar{D} \times \bar{D}$ such that

- $G(z, \xi) = G(\xi, z)$ and $G(z, \xi') = 0$ for any $z \in D$ and $\xi' \in \partial D$;
- the function $g(z, \xi)$ is continuous on $D \times D$ and continuous with respect to ξ on $\bar{D}$ for every $z \in D$;
- the function $g(z, \xi)$ is harmonic with respect to z for every $\xi \in D$ and harmonic with respect to ξ for every $z \in D$.

Exercise 4.4 Show that for every domain $D \subset \bar{\mathbb{C}}$ there exists at most one Green's function on D.

For a connected, simply connected domain D bounded by Jordan curves, a Green's function does exist and can be expressed in terms of a biholomorphic map from D onto the unit disk $\Lambda = \{z \in \mathbb{C} \mid |z| < 1\}$, which exists by the Riemann mapping theorem.

Theorem 4.3 *Let $D \subset \mathbb{C}$ be a connected, simply connected domain and $w \colon D \to \Lambda$ be a biholomorphic map onto the unit disk. Set*

$$W(z, \xi) = \frac{w(z) - w(\xi)}{1 - w(z)\overline{w(\xi)}}.$$

Then $G(z, \xi) = G(\xi, z) = \frac{1}{2\pi} \ln|W(z, \xi)|$ is a Green's function for the domain D.

Proof Let us fix an arbitrary point $\xi \in D$ and regard $w_\xi(z) = W(z, \xi)$ as a function of $z \in D$. This function is a conformal map from the domain $D = \{z\}$ onto the disk Λ which sends the point $z = \xi$ to 0. In particular,

$$w_\xi(z) = w_\xi(\xi) + (z - \xi)\, w'_\xi(\xi) + (z - \xi)\, o(1),$$

where $w_\xi(\xi) = 0$ and $w'_\xi(\xi) \neq 0$.

Thus, the function $\frac{w_\xi(z)}{z - \xi}$ is holomorphic and does not vanish. Therefore, the function $\ln \frac{w_\xi(z)}{z - \xi}$ is holomorphic and the function

$$g(z, \xi) = G(z, \xi) - \frac{1}{2\pi} \ln|z - \xi| = \frac{1}{2\pi} \ln\left|\frac{W(z, \xi)}{z - \xi}\right|$$

$$= \frac{1}{2\pi} \ln\left|\frac{w_\xi(z)}{z - \xi}\right| = \frac{1}{2\pi} \Re \ln \frac{w_\xi(z)}{z - \xi}$$

is harmonic with respect to z for every $\xi \in D$. Since $w_\xi(z) = W(z, \xi) = w_z(\xi)$, this function is symmetric and continuous with respect to the pair of variables

(z, ξ) on $D \times D$. Besides, by Carathéodory's theorem 3.10, the function $w_\xi(z)$ is continuous on the closure $\bar{D}$ and $|w_\xi(\partial D)| = 1$, whence $G(\partial D, \xi) = 0$.

The theorem proved above allows one to efficiently calculate Green's functions for simple domains. In particular,

- for the unit disk Λ, the map $w_\xi(z) = \frac{z-\xi}{1-z\bar{\xi}}$ produces the Green's function

$$G_\Lambda(z, \xi) = \frac{1}{2\pi} \ln \frac{|z - \xi|}{|1 - z\bar{\xi}|};$$

- for the right half-plane $H = \{z \in \mathbb{C} \mid \Re z > 0\}$, the map $w_\xi(z) = \frac{z-\xi}{z+\xi}$ produces the Green's function

$$G_H(z, \xi) = \frac{1}{2\pi} \ln \left| \frac{z - \xi}{z + \xi} \right|.$$

4.4 Dirichlet Problem

Definition 4.3 The *Dirichlet problem* for a domain D consists in finding a function u, harmonic on D and continuous on the closure $\bar{D}$, with prescribed (bounded, continuous) boundary values $u|_{\partial D} = \varphi$ on ∂D.

Exercise 4.5 Show that the Dirichlet problem has at most one solution.

The Green's function solves the Dirichlet problem in the following sense.

Theorem 4.4 *Let* $G(z, \xi) = G_D(z, \xi)$ *be the Green's function for a domain* D. *Then the function*

$$u(z) = \oint_{\xi \in \partial D} \varphi(\xi) \frac{\partial}{\partial n} G(z, \xi) \, ds$$

is a solution to the Dirichlet problem for D.

Proof The function

$$u(z) = \oint_{\xi \in \partial D} \varphi(\xi) \frac{\partial}{\partial n} G(z, \xi) \, ds$$

is harmonic, because the function $G(z, \xi)$ is harmonic with respect to z for every $\xi \in D$. It remains to show that the harmonic function $u(z)$ satisfies the relation

$$u(z) = \oint_{\xi \in \partial D} u(\xi) \frac{\partial}{\partial n} G(z, \xi) \, ds.$$

By our definitions,

$$G(z, \xi) = \frac{\ln r}{2\pi} + g(z, \xi),$$

where $r = |z - \xi|$. By Theorem 4.2, we have

$$2\pi u(z) = \oint_{\xi \in \partial D} \left\{ u(\xi) \frac{\partial \ln r}{\partial n} - \frac{\partial u}{\partial n}(\xi) \ln r \right\} ds$$

and

$$\oint_{\xi \in \partial D} \left\{ u(\xi) \frac{\partial g}{\partial n}(z, \xi) - g(z, \xi) \frac{\partial u}{\partial n}(\xi) \right\} ds = 0$$

for $z \in D$. Therefore,

$$\oint_{\xi \in \partial D} \left\{ u(\xi) \frac{\partial G}{\partial n}(z, \xi) - G(z, \xi) \frac{\partial u}{\partial n}(\xi) \right\} ds$$

$$= \oint_{\xi \in \partial D} \left\{ u(\xi) \frac{\partial (\frac{\ln r}{2\pi} + g(z, \xi))}{\partial n} - \left(\frac{\ln r}{2\pi} + g(z, \xi) \right) \frac{\partial u}{\partial n}(\xi) \right\} ds = u(z).$$

Besides, $G(z, \xi) = 0$ for $\xi \in \partial D$.

The Green's functions found above for simple domains allow one to solve the Dirichlet problem for these domains. As we know, the Green's function for the disk Λ is equal to $G_\Lambda(z, \xi) = \frac{1}{2\pi} \ln \frac{|z-\xi|}{|1-z\bar{\xi}|}$. Set $z = re^{i\varphi}$, $\xi = \rho e^{i\theta}$. Then

$$\frac{\partial}{\partial n} G(re^{i\varphi}, \rho e^{i\theta}) = \frac{1}{2\pi} \Re \frac{\partial}{\partial \rho} \ln \frac{re^{i\varphi} - \rho e^{i\theta}}{1 - r\rho e^{i(\varphi - \theta)}} \Big|_{\rho=1}$$

$$= \frac{1}{2\pi} \Re \left\{ \frac{e^{i\theta}}{e^{i\theta} - re^{i\varphi}} + \frac{re^{i(\theta-\varphi)}}{1 - re^{i(\theta-\varphi)}} \right\} = \frac{1}{2\pi} \frac{1 - r^2}{1 + r^2 - 2r \cos(\theta - \varphi)}.$$

Thus, the solution of the Dirichlet problem for Λ is given by the *Poisson integral formula for the disk* $|z| < 1$:

$$u(re^{i\varphi}) = \frac{1}{2\pi} \int_0^{2\pi} \frac{1 - r^2}{1 + r^2 - 2r \cos(\theta - \varphi)} u(e^{i\theta}) \, d\theta.$$

Exercise 4.6 Prove the *Poisson integral formula for the right half-plane* ($\Re z > 0$):

$$u(x + iy) = \frac{x}{\pi} \int\limits_{-\infty}^{\infty} \frac{u(i\eta)\,d\eta}{(y - \eta)^2 + x^2}.$$

Exercise 4.7 Prove the *Schwarz integral formula* for the disk

$$F(z) = \frac{1}{2\pi} \int\limits_{0}^{2\pi} \frac{Re^{i\theta} + z}{Re^{i\theta} - z} \Re F(Re^{i\theta})\,d\theta + iC \quad (|z| < R)$$

and for the right half-plane

$$F(z) = \frac{x}{\pi} \int\limits_{-\infty}^{\infty} \frac{\Re F(i\eta)}{(i\eta - z)}\,d\eta + iC,$$

which allow one to recover a holomorphic function from the boundary values of its real part.

Chapter 5
Riemann Surfaces and Their Modules

5.1 Riemann Surfaces

A *Riemann surface* is a one-dimensional complex manifold. A Riemann surface is defined as an equivalence class of atlases of charts on a surface with biholomorphic transition maps.

Definition 5.1 Let P be a topological surface.

- A *chart* on P is a pair $(U_\alpha, \varphi_\alpha)$ where $U_\alpha \subset P$ and $\varphi_\alpha : U_\alpha \to \mathbb{C}$ is a homeomorphism onto an open simply connected subset in $\mathbb{C}$.
- A *holomorphic atlas* on P is a family of charts $\{(U_\alpha, \varphi_\alpha)\}$ such that

$$P = \bigcup_\alpha U_\alpha \quad \text{and} \quad \varphi_\beta \varphi_\alpha^{-1}|_{\varphi_\alpha(U_\alpha \cap U_\beta)} : \varphi_\alpha(U_\alpha \cap U_\beta) \to \varphi_\beta(U_\alpha \cap U_\beta)$$

  are holomorphic functions.
- Holomorphic atlases on P are said to be *equivalent* if their union is again a holomorphic atlas.
- An equivalence class of holomorphic atlases on P is called a *complex structure*. A surface endowed with a complex structure is called a *Riemann surface*.
- A map $F : P \to \mathbb{C}$ is said to be *holomorphic* if the function $F\varphi_\alpha^{-1} : \varphi_\alpha(U_\alpha) \to \mathbb{C}$ is holomorphic for every chart $(U_\alpha, \varphi_\alpha)$ of the holomorphic atlas.

Exercise 5.1 Show that an arbitrary domain $U \subset \bar{\mathbb{C}}$ is a Riemann surface.

To complete the description of the category of Riemann surfaces, it remains to describe morphisms between them. These are holomorphic maps in the following sense.

© Springer Nature Switzerland AG 2019
S. M. Natanzon, *Complex Analysis, Riemann Surfaces and Integrable Systems*,
Moscow Lectures 3, https://doi.org/10.1007/978-3-030-34640-9_5

Definition 5.2 Let P and Q be Riemann surfaces defined by holomorphic atlases $\{(U_\alpha, \varphi_\alpha)\}$ and $\{(V_\beta, \psi_\beta)\}$, respectively. A map $f: P \to Q$ is said to be *holomorphic* if the functions

$$\psi_\beta f \varphi_\alpha^{-1}: \varphi_\alpha(U_\alpha \cap f^{-1}(V_\beta)) \to \psi_\beta(V_\beta)$$

are holomorphic for $U_\alpha \cap f^{-1}(V_\beta) \neq \varnothing$.

Thus, meromorphic functions on a Riemann surface P are holomorphic morphisms from P to the Riemann sphere.

Exercise 5.2 Show that the definition of a holomorphic map does not depend on the choice of a holomorphic atlas in the equivalence class defining the Riemann surface.

Invertible holomorphic maps between Riemann surfaces are called (biholomorphic) *isomorphisms*.

5.2 Riemann Surfaces of Analytic Functions

Riemann surfaces play a key role in the study of global properties of holomorphic functions. The main methodological problem arising here is that many important functions are "multivalued," i.e., are not maps between sets. The reason is that the analytic continuation of a canonical element of a function along a closed contour may produce a new canonical element.

For example, the analytic continuation of a canonical element (U_a, f_a) of the function $f(z) = \sqrt{z}$ along a closed contour around 0 gives the canonical element $(U_a, -f_a)$ of the same function. A similar property holds for the function $f(z) = \ln(z)$ and many other functions important for natural sciences, in particular, those defined implicitly by equations of the form $F(x, y) = 0$. This suggests to consider the set of all canonical elements that can be obtained from one of them by analytic continuation.

Definition 5.3 An *analytic function* F is the set of all analytic continuations $\{(U^l, f^l)\}$ of a canonical element (U_a, f_a) along all paths l starting at a. Canonical elements (U_a, f_a) and (V_b, g_b) are *equivalent* if $g_a = f_b$ on $U_a \cap V_b$. Two analytic functions are *equal* if they have at least one pair of equivalent canonical elements.

An analytic function $F = \{(U^l, f^l)\}$ generates a Riemann surface P_F as follows. Let $\varphi^l: U^l \to \mathbb{C}$ be the "tautological" map identifying U^l with the corresponding domain in $\mathbb{C}$. Let us identify points $p^s \in U^s$ and $p^t \in U^t$ if they have neighborhoods $V^s \subset U^s$ and $V^t \subset U^t$, respectively, such that for any pair of points $q^s \in V^s, q^t \in V^t$ satisfying $\varphi^s(q^s) = \varphi^t(q^t)$, we have $f^s(q^s) = f^t(q^t)$.

On the resulting set P_F, the charts $\{(U^l, f^l)\}$ define the structure of a Riemann surface, which is called the *Riemann surface of the analytic function* F. The functions $\{f^l\}$ generate a holomorphic map $f_F: P_F \to \mathbb{C}$. Thus, an analytic function F is equivalent to the holomorphic function f_F in the ordinary sense, but defined on the Riemann surface P_F rather than on the complex plane $\mathbb{C}$.

Exercise 5.3 Construct the Riemann surfaces of the functions $w = \ln(z)$ and

$$w^2 = z^n + az^{n-1} + \ldots + bz + c.$$

Exercise 5.4 Give a definition of a meromorphic analytic function F, its Riemann surface, and a map $f_F : P_F \to \bar{\mathbb{C}}$. Show that the map $f_F : P_F \to \bar{\mathbb{C}}$ generated by a meromorphic analytic function F is a meromorphic function on its Riemann surface.

5.3 Uniformization

Now we turn to the classification of Riemann surfaces, starting with the case of simply connected ones. In this case, there is a remarkable uniformization theorem [26, Chap. 9], which was announced by Riemann but proved only 50 years later by P. Koebe.

Theorem 5.1 *Every simply connected Riemann surface is isomorphic to* $\bar{\mathbb{C}}$, $\mathbb{C}$, *or* $\Lambda = \{z \in \mathbb{C} \mid |z| < 1\}$.

For subsets of the Riemann sphere $\bar{\mathbb{C}}$, this follows from the Riemann mapping theorem. But the general case is much more difficult.

Now let P be an arbitrary Riemann surface. Consider its topological universal simply connected covering $\psi : \tilde{P} \to P$. It defines a representation of P as the quotient surface $\tilde{P}/\Gamma$ where Γ is a discrete group acting on $\tilde{P}$ without fixed points. A holomorphic atlas on P generates a holomorphic atlas on $\tilde{P}$ such that ψ is a holomorphic map. Equivalent atlases on P generate equivalent atlases on $\tilde{P}$. Thus, the map ψ determines a Riemann surface structure on $\tilde{P}$ with respect to which ψ is a holomorphic map and $\Gamma \in \operatorname{Aut} \tilde{P}$ (prove this).

Exercise 5.5 Let Λ_F be a simply connected covering of the Riemann surface P_F of an analytic function F. Consider the covering projection $\psi : \Lambda_F \to P_F$, and let $f_\Lambda = f_F \psi$. Show that if $\operatorname{Aut} P_F = 1$, then the Riemann surface P_F is biholomorphically equivalent to the Riemann surface Λ_F/Γ where $\Gamma = \{g \in \operatorname{Aut} \Lambda \mid f_\Lambda g = f_\Lambda\}$.

The isomorphism problem for quotient surfaces is solved as follows.

Exercise 5.6 Show that quotient surfaces of nonisomorphic simply connected Riemann surfaces are not isomorphic, and that groups $\Gamma_1, \Gamma_2 \subset \operatorname{Aut} \tilde{P}$ acting without fixed points generate isomorphic Riemann surfaces $\tilde{P}/\Gamma_1$, $\tilde{P}/\Gamma_2$ if and only if Γ_1 and Γ_2 are conjugate in the group $\operatorname{Aut} \tilde{P}$, i.e., $\Gamma_1 = A\Gamma_2 A^{-1}$ for some $A \in \operatorname{Aut} \tilde{P}$.

The uniformization theorem and Exercise 5.6 reduce the problem of describing the Riemann surfaces to that of describing the conjugacy classes of discrete groups acting without fixed points on simply connected uniformizing surfaces: the Riemann sphere, the complex plane, and the unit disk.

Exercise 5.7 Show that every automorphism of the Riemann sphere has a fixed point. Show that every discrete group of automorphisms of the complex plane acting without fixed points is generated by one or two translations. Show that the quotient $\mathbb{C}/\Gamma$ by the group Γ generated by an arbitrary translation is isomorphic to $\mathbb{C} \setminus 0$.

Definition 5.4 A discrete subgroup Γ of the group of automorphisms of the unit disk Λ or the upper half-plane H is called a *Fuchsian group*.

Exercise 5.8 Show that every Fuchsian group either is generated by a single element, or is noncommutative.

Example 5.1 The following group, called the *classical modular group*, is Fuchsian:

$$\text{Mod} = \left\{ z \mapsto \frac{az+b}{cz+d} \mid a, b, c, d \in \mathbb{Z}, \ ad - bc = 1 \right\} \cong \text{PSL}(2, \mathbb{Z}).$$

Exercise 5.9 Find simple generators and a fundamental domain of the modular group. (*Hint.* Consider the domain shown in Fig. 5.1.)

We conclude this section with an intermediate summary.

Theorem 5.2 *Every Riemann surface P is isomorphic to exactly one Riemann surface from the following list: $\check{\mathbb{C}}$, $\mathbb{C}$, $\mathbb{C} \setminus 0$, the torus $\mathbb{C}/\Gamma'$ where Γ' is the group generated by two noncollinear translations, and Λ/Γ where $\Gamma \subset \text{Aut}\,\Lambda \cong \pi_1(P, p)$ is a Fuchsian group acting without fixed points.*

Fig. 5.1

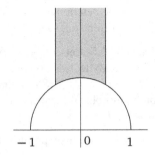

-1 0 1

5.4 Moduli of Compact Riemann Surfaces of Genus 1

The set of isomorphism classes of Riemann surfaces of a given topological type is called the *moduli space.*

We know already that the only compact Riemann surface of genus 0 is the Riemann sphere.

Theorem 5.3 *The moduli space M_1 of tori (compact Riemann surfaces of genus 1) can be naturally identified with the space H/Mod.*

Proof Let P be a torus. Then, by Theorem 5.2, we have $P = C/\Gamma$ where Γ is generated by translations by vectors $\sigma, \tau \in \mathbb{C}$ with $\tau/\sigma \in \mathbb{C} \setminus (\mathbb{R} \cup \infty)$. By Exercise 5.6, we may assume that $\sigma = 1$ and $\tau \in H$. Moreover, by Exercise 5.6, two pairs of vectors $(1, \tau)$ and $(1, \tau')$, where $\tau, \tau' \in H$, generate isomorphic Riemann surfaces if and only if they generate groups of translations that are conjugate in $\mathrm{Aut}\,\mathbb{C}$. This means that for some $A \in \mathbb{C}$ the generators $(A, A\tau')$ of the group of translations can be expressed in terms of the generators $(1, \tau)$ as $A = c\tau + d$ and $A\tau' = a\tau + b$ where $\begin{pmatrix} a & b \\ c & d \end{pmatrix} \in \mathrm{SL}(2, \mathbb{Z})$. Thus, $\tau' = \frac{a\tau+b}{c\tau+d} = \gamma\tau$ where $\gamma \in \mathrm{Mod}$.

Exercise 5.10 The moduli space of compact Riemann surfaces of genus 1 is endowed with a natural complex structure and is isomorphic to the complex plane.

5.5 Automorphisms of the Upper Half-plane

Instead of the unit disk Λ, it is sometimes convenient to consider the upper half-plane $H = \{z \in \mathbb{C} \mid \Im z > 0\}$, which is isomorphic to it. The reason is that the automorphism group $\mathrm{Aut}\,H$ of H has a simple form:

$$\mathrm{Aut}\,H = \left\{ Az = \frac{az+b}{cz+d} \mid a, b, c, d \in \mathbb{R}, \ ad - bc > 0 \right\} \cong \mathrm{PSL}(2, \mathbb{R}).$$

Exercise 5.11 Show that $\mathrm{Aut}\,H$ coincides with the group of isometries of the metric $ds = \frac{|dz|}{\Im z}$, which turns H into the *Poincaré model of Lobachevsky (hyperbolic) geometry*. Find the straight lines (i.e., geodesics) of Lobachevsky geometry in this model.

This metric of constant negative curvature is called *hyperbolic*. It can be carried over to the quotient of H by a discrete group acting without fixed points. That is why, all Riemann surfaces uniformized by the half-plane H are called *hyperbolic*.

The fixed points of an automorphism $Az = \frac{az+b}{cz+d}$ are the roots z_1, z_2 of the equation $cz^2 + (d - a)z - b = 0$. The automorphism A is said to be

- *elliptic* if $z_1, z_2 \notin \mathbb{R}$; in this case, $z_2 = \bar{z}_1$ and A has one fixed point in H (*example*: $A(z) = \frac{z+1}{-z+1}$).

Fig. 5.2

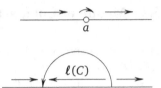

Fig. 5.3

- *parabolic* if $z_1 = z_2$; in this case, $z_1 = z_2 \in \mathbb{R}$ and A has no fixed points in H and only one fixed point in $\mathbb{R} \cup \infty$ (*example:* $A(z) = z + b$).
- *hyperbolic* if $z_1 \neq z_2 \in \mathbb{R}$; in this case, A has no fixed points in H, but exactly two fixed points in $\mathbb{R} \cup \infty$ (*example:* $A(z) = \lambda z$).

Exercise 5.12 Show that every parabolic automorphism C with fixed point $a \in \mathbb{R}$ is conjugate in the group Aut H to an automorphism of the form $z \mapsto z + \lambda$ and has the form $C(z) = \frac{(1-a\gamma)z+a^2\gamma}{-\gamma z+(1+a\gamma)}$; moreover, if $\gamma > 0$ and $r \in \mathbb{R} \setminus a$, then $C(r) > r$.

The action of such an automorphism in a neighborhood of the point a is shown in Fig. 5.2. The fixed point $a = a(C)$ of a parabolic automorphism C will also be denoted by $\alpha(C) = \beta(C)$.

Exercise 5.13 Show that every hyperbolic automorphism C is conjugate in the group Aut H to an automorphism of the form $z \mapsto \lambda z$, $\lambda > 0$, and has the form $Cz = \frac{(\lambda a-\beta)z+(1-\lambda)\alpha\beta}{(\lambda-1)z+(\alpha-\lambda\beta)}$. The number $\lambda > 1$ is called the *shift parameter of* C, while the points $\alpha = \alpha(C)$ and $\beta = \beta(C)$ are called the *attracting* and *repelling fixed points* of C, respectively. The semicircle $\ell(C)$ connecting these points (see Fig. 5.3) is invariant under C (the invariant line of Lobachevsky geometry).

5.6   Types of Riemann Surfaces

A Riemann surface uniformized by the half-plane H and homeomorphic to a cylinder is isomorphic to $H/\langle C \rangle$ where $\langle C \rangle$ is the group generated by one parabolic or hyperbolic automorphism C acting without fixed points. The metric of constant negative curvature on H generates a metric of constant negative curvature on $H/\langle C \rangle$. In both cases, the surface $H/\langle C \rangle$ is homeomorphic to a cylinder, but the resulting Riemann surfaces are not isomorphic, since a parabolic automorphism is not conjugate in the group Aut H to a hyperbolic automorphism.

Exercise 5.14 Show that the Riemann surface $H/\langle C \rangle$ generated by a parabolic automorphism C is isomorphic to $\Lambda \setminus 0$ and has no closed geodesics.

Exercise 5.15 Show that the Riemann surface $H/\langle C \rangle$ generated by a hyperbolic automorphism C has a unique geodesic, which coincides with the image of the invariant line of C, and this geodesic is a simple not null-homotopic contour of minimal length. Show that such surfaces are isomorphic if and only if the corresponding minimal lengths coincide.

Thus, there are exactly three different classes of complex structures on the cylinder. One of them, the *Riemann sphere with two punctures* $\bar{\mathbb{C}} \setminus \{\infty \cup 0\} = \mathbb{C} \setminus 0$ (see Sect. 5.3), will also be called a surface of type $(0, 0, 2)$. The second one, generated by a parabolic element and isomorphic to the *punctured disk* $\Lambda \setminus 0$, will also be called a surface of type $(0, 1, 1)$. The third one, generated by a parabolic element, will be called a *disk with a hole*, or a Riemann surface of type $(0, 2, 0)$.

In what follows, we will consider only Riemann surfaces with finitely generated fundamental groups. Such a surface is homeomorphic to a compact surface of genus g with n connected, simply connected, pairwise disjoint, closed subsets removed. A neighborhood of a removed subset is a hyperbolic Riemann surface homeomorphic to a cylinder. Therefore, it is either a punctured disk or a disk with a hole. Accordingly, the removed subset will be called a *puncture* or a *hole*.

Thus, with a Riemann surface with finitely generated fundamental group we can associate the *genus g*, the *number of holes k*, and the *number of punctures m*. The triple (g, k, m) will be called the *type of the surface*. We have already explained the meaning of this definition for Riemann cylinders.

Exercise 5.16 Show that the number of punctures of a hyperbolic Riemann surface H/Γ coincides with the number of conjugacy classes of parabolic elements in the group Γ.

The set $M_{g,k,m}$ of isomorphism classes of hyperbolic Riemann surfaces of type (g, k, m) is called the *moduli space of Riemann surfaces of type (g, k, m)*. Below we will show that the set $M_{g,k,m}$ has a natural structure of a topological space and will study its topology.

Exercise 5.17 Verify that

- each of the spaces $M_{0,0,0}$, $M_{0,0,1}$, $M_{0,1,0}$, $M_{0,1,1}$, and $M_{0,0,2}$ consists of a single element ($\bar{\mathbb{C}}$, $\mathbb{C}$, Λ, $\Lambda \setminus 0$, and $\mathbb{C} \setminus 0$, respectively);
- the space $M_{0,2,0}$ is homeomorphic to $\mathbb{R}$;
- the space $M_{1,0,0} = M_1$ is homeomorphic to H/Mod.

It remains to consider the moduli spaces $M_{g,k,m}$ where $6g + 3k + 2m > 6$. In what follows, we consider only surfaces of type (g, k, m) with $6g + 3k + 2m > 6$. All of them are hyperbolic.

5.7 Sequential Sets of Automorphisms

We turn to the study of hyperbolic surfaces uniformized by noncommutative Fuchsian groups.

For automorphisms $C_1, C_2 \in \mathrm{Aut}\, H$ with finite fixed points in $\mathbb{R} \subset \mathbb{C}$, we set $C_1 < C_2$ if $\alpha(C_1) \leq \beta(C_1) < \alpha(C_2) \leq \beta(C_2)$.

A set $\{C_1, C_2, C_3\} \in \mathrm{Aut}\, H$ is said to be *sequential* if it contains no elliptic automorphisms, $C_1 \cdot C_2 \cdot C_3 = 1$, and there exists an automorphism $D \in \mathrm{Aut}\, H$

such that the automorphisms $\tilde{C}_i = DC_i D^{-1}$ ($i = 1, 2, 3$) have finite fixed points and $\tilde{C}_1 < \tilde{C}_2 < \tilde{C}_3$.

In geometric arguments, it is convenient to proceed from the upper half-plane H to the disk $\Lambda = \{z \in \mathbb{C} \mid |z| < 1\}$. The group Aut Λ of holomorphic automorphisms of Λ also splits into the sets of elliptic, parabolic, and hyperbolic automorphisms, which have 0, 1, 2 fixed points, respectively, on the absolute $\partial \Lambda = \{z \in \mathbb{C} \mid |z| = 1\}$.

A biholomorphic isomorphism $\varphi \colon H \to \Lambda$ sends the metric on H to a metric on Λ invariant under Aut Λ. Straight lines (i.e., geodesics) in this metric are circular arcs orthogonal to $\partial \Lambda$.

The isomorphism φ sends

- a hyperbolic automorphism $C \in$ Aut H with invariant line $\ell(C)$ to the automorphism $\varphi C \varphi^{-1} \in$ Aut Λ with invariant line $\ell(\varphi(C))$;
- a parabolic automorphism $C \in$ Aut H with fixed point γ_C to the automorphism $\varphi C \varphi^{-1} \in$ Aut Λ with fixed point $\varphi(\gamma_C)$.

Lemma 5.1 *Let $\{C_1, C_2, C_3\}$ be a sequential set. Then the set $\{C_1 C_2 C_1^{-1}, C_1, C_3\}$ is also sequential.*

Proof The fixed points of the translations C_i divide the absolute into 6 arcs γ_i shown in Fig. 5.4.

Let α and β be the attracting and repelling fixed points of the translation $C_1 C_2 C_1^{-1}$. Then $\ell(C_1 C_2 C_1^{-1}) = C_1 \ell(C_2)$, and hence $\beta \in \gamma_4 \cup \gamma_5 \cup \gamma_6$. On the other hand, $C_1 \ell(C_2) = C_3^{-1} C_2^{-1} \ell(C_2) = C_3^{-1} \ell(C_2)$, and hence $\beta \in \gamma_3 \cup \gamma_2 \cup \gamma_1 \cup \gamma_6$. Thus, $\beta \in \gamma_6$. In a similar way, $\alpha \in \gamma_6$. This means that the set $\{C_1 C_2 C_1^{-1}, C_1, C_3\}$ is also sequential.

A set $\{C_1, \ldots, C_n\}$ consisting of hyperbolic automorphisms $C_1, \ldots, C_k$ and parabolic automorphisms $C_{k+1}, \ldots, C_n$ will be called a *sequential set of type* $(0, k, n - k)$ if the sets $\{C_1 \ldots C_{i-1}, C_i, C_{i+1} \ldots C_n\}$ are sequential for all $i = 1, \ldots, n - 1$.

Exercise 5.18 Let $\{C_1, \ldots, C_n\} \in$ Aut Λ be a sequential set. Show that the semicircle $\ell = C_j C_{j+1} \ldots C_{n-1} C_n(\ell(C_1))$ lies between $\ell(C_{j-1})$ and $\ell(C_j)$.

Fig. 5.4

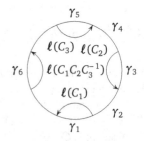

Below, we will consider only Fuchsian groups without elliptic elements. So, in what follows, by a *Fuchsian group* we mean a discrete subgroup of Aut Λ or Aut H in which all nontrivial elements act on Λ or H without fixed points (i.e., are hyperbolic or parabolic).

Lemma 5.2 *A sequential set* $V = \{C_1, \ldots, C_n\} \subset$ Aut Λ *of type* $(0, k, m)$ *generates a Fuchsian group* Γ. *In this case,* Λ/Γ *is a sphere with* k *holes and* m *punctures.*

Proof First, let $k > 0$. Consider a point $O_1 \subset \ell(C_1)$ and put $O_i = C_i C_{i+1} \ldots C_n(O_1)$.

Let r_i be a geodesic ray intersecting $\ell(C_i)$ that starts at O_i and ends at a point of the absolute. Then $d_i = C_i^{-1} r_i$ is a geodesic ray that ends at O_{i+1}. The rays $\{r_i, d_i \ (i = 1, \ldots, n)\}$ bound a noncompact domain M.

Each arc $\ell(C_i)$ cuts off from M a "tail" M_i that reaches the boundary $\partial \Lambda$. The relative position of the images $C_j(M_i)$ of these tails is determined by the relative position of the lines $C_j(\ell(C_i))$. Therefore, according to Exercise 5.18, the images $C_1(M_i)$ meet $\partial \Lambda$ at the segment between $\alpha(C_1)$ and $\beta(C_n)$. For $j > 1$, the images $C_j(M_i)$ meet $\partial \Lambda$ at the segment between $\alpha(C_j)$ and $\beta(C_{j-1})$.

On the other hand, the relative position of the images $\gamma(M_i)$ and the tails M_i determines the relative position of the domains $\gamma(M)$ and M for $\gamma \in$ Aut Λ. Therefore (see Fig. 5.5), the sequence of polygons

$$M, \ C_1 M, \ C_1 C_2 M, \ \ldots, \ C_1 \ldots C_{n-1} M$$

makes a simple circuit around the point O_1, and

$$C_1 \ldots C_j \bar{M} \cap \bar{M} = O_1 \quad \text{for } j < n.$$

Thus, $\Lambda = \bigcup_{\gamma \in \Gamma} \gamma \bar{M}$, and the domains $\gamma_1 M \cap \gamma_2 M$ are nonempty for $\gamma_1 \neq \gamma_2$. It follows that $\bar{M}$ is a fundamental domain for the discrete group Γ generated by the automorphisms $\{C_1, \ldots, C_n\}$ and acting without fixed points.

Fig. 5.5

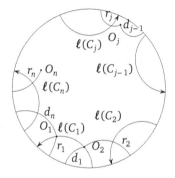

Fig. 5.6

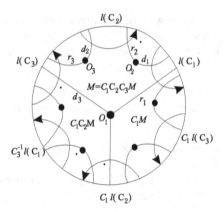

The automorphism C_i identifies r_i and d_i, so on the surface Λ/Γ there appears a hole for $i \leq k$ and a puncture for $i > k$. This construction works also in the case $k = 0$ if for O_1 one takes a point sufficiently close to the fixed point of the parabolic automorphism C_1.

Example 5.2 Figure 5.6 shows a circuit around the vertex O_1 for $k = 3$.

By a *sequential set of type* (g, k, m) we mean a set

$$\{A_1, B_1, \ldots, A_g, B_g, C_1, \ldots, C_n\}$$

such that A_i, B_i $(i = 1, \ldots, g)$ are hyperbolic automorphisms and

$$\{A_1, B_1 A_1^{-1} B_1^{-1}, \ldots, A_g, B_g A_g^{-1} B_g^{-1}, C_1, \ldots, C_n\}$$

is a sequential set of type $(0, 2g + k, m)$.

We say that a Fuchsian group $\Gamma \subset \operatorname{Aut} \Lambda$ (or $\Gamma \subset \operatorname{Aut} H$) is a *Fuchsian group of type* (g, k, m) if Λ/Γ (respectively, H/Γ) belongs to $M_{g,k,m}$.

Theorem 5.4 *A sequential set of type* (g, k, m) *generates a Fuchsian group* Γ *of type* (g, k, m).

Proof Let $\{A_1, B_1, \ldots, A_g, B_g, C_{g+1}, \ldots, C_n\} \in \operatorname{Aut} \Lambda$ be a sequential set of type (g, m, k). For $g = 0$, the claim follows from Lemma 5.2. Let $g > 0$. Put $C_i = [A_i B_i]$ $(i = 1, \ldots, g)$. Our definitions ensure that the geodesics $\ell(A_i)$, $\ell(B_i)$, $\ell(C_i)$ are located as shown in Fig. 5.7.

Let $O_1 \in \ell(C_1)$, and let M be the polygon constructed in the proof of Lemma 5.2. For $i \leq g$, replace the rays r_i, d_i by the geodesic segments with the vertices

$$O_i,\ A_i B_i^{-1} A_i^{-1} O_i,\ B_i^{-1} A_i^{-1} O_i,\ A_i^{-1} O_i,\ B_i A_i B_i^{-1} A_i^{-1} O_i = O_{i+1}.$$

Fig. 5.7

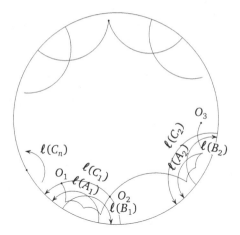

The result is a new polygon $\tilde{M}$ (see Fig. 5.7). Using the same arguments as in the proof of Lemma 5.2, we conclude that the sequence of polygons

$$\tilde{M}, \ A_1\tilde{M}, \ A_1B_1\tilde{M}, \ A_1B_1A_1^{-1}\tilde{M}, \ C_1\tilde{M}, \ C_1A_2\tilde{M}, \ C_1A_2B_2\tilde{M},$$

$$C_1A_2B_2A_2^{-1}\tilde{M}, \ C_1C_2\tilde{M}, \ \ldots, \ C_1C_2\ldots C_{n-1}M$$

makes a circuit around the point O_1 and, therefore, the set $\{A_1, \ldots, C_n\}$ generates a Fuchsian group Γ. It is not difficult to see that each pair (A_i, B_i) generates a "handle" of the surface Λ/Γ. Hence, Λ/Γ is a surface of type (g, k, m).

5.8 The Geometry of Fuchsian Groups

Let P be a surface of type (g, k, m). A set

$$v = \{a_i, \ b_i \ (i = 1, \ldots, g), \ c_i \ (i = g+1, \ldots, n)\}$$

of generators of the group $\pi_1(P, p)$ is said to be *standard* if v generates $\pi_1(P, p)$ with the defining relation

$$\prod_{i=1}^{g}[a_i, b_i] \prod_{i=g+1}^{n} c_i = 1$$

and can be represented by a set of simple contours

$$\tilde{v} = \{\tilde{a}_i, \ \tilde{b}_i \ (i = 1, \ldots, g), \ \tilde{c}_i \ (i = g+1, \ldots, n)\}$$

Fig. 5.8

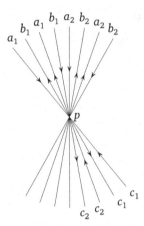

Fig. 5.9

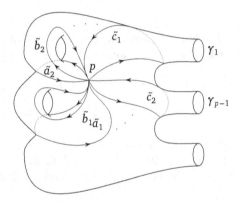

satisfying the following properties:

- the contour $\tilde{c}_i$ is homologous to zero and cuts off from the surface P a single hole for $i \leq g + k$ and a single puncture for $i > g + k$;
- $\tilde{a}_i \cap \tilde{b}_j = \tilde{a}_i \cap \tilde{c}_j = \tilde{b}_i \cap \tilde{c}_j = \tilde{c}_i \cap \tilde{c}_j = p$;
- in a neighborhood of the point p, the contours $\tilde{v}$ are located as shown in Fig. 5.8.

In this case, the set of contours $\tilde{v}$ is located on P as shown in Fig. 5.9.

Now let $\Gamma \subset \operatorname{Aut} \Lambda$ be a Fuchsian group, $P = \Lambda/\Gamma$ be the corresponding Riemann surface, $\Phi: \Lambda \to P$ be the natural projection, and $q \in \Phi^{-1}(p)$. With an automorphism $C \in \Gamma$ we associate an oriented geodesic segment $\ell_q(C) \subset \Lambda$ that starts at q and ends at $C(q)$. The correspondence $C \mapsto \Phi(\ell_q(C))$ generates an isomorphism $\Phi_q: \Gamma \to \pi_1(P, p)$.

Lemma 5.3 *Let* $V = \left\{ A_i, \ B_i \ (i = 1, \ldots, g), \ C_i \ (i = g + 1, \ldots, n) \right\}$ *be a sequential set of type* (g, k, m), Γ *be the Fuchsian group generated by this set, and* $P = \Lambda/\Gamma$. *Then* $v_q = \Phi_q(V)$ *is a standard set of generators of the group* $\pi_1(P, p)$.

Proof Consider the fundamental domain M constructed in the proof of Lemma 5.2 and Theorem 5.4. For $i > g$, connect the points O_i and O_{i+1} by pairwise disjoint segments $c_i \subset M$. (Here $O_{n+1} = O_1$.) On H consider the geodesic segments

$$a_i b_i^{-1} a_i^{-1} = [O_i, A_i B_i^{-1} A_i^{-1} O_i], \quad 5 a_i^{-1} = [A_i B_i^{-1} A_i^{-1} O_i, B_i^{-1} A_i^{-1} O_i],$$

$$c_i^{-1} = [O_i, C^{-1} O_i].$$

Then the natural projection $\Phi : \Lambda \to P$ generates a standard set of generators

$$v_{O_1} = \{\Phi(a_i), \ \Phi(b_i) \ (i = 1, \dots, g), \ \Phi(c_i) \ (i = g + 1, \dots, n)\} \in \pi_1(P, \Phi(O_1)).$$

If we continuously move the point O_1 to q, the set v_{O_1} turns into the standard set of generators v_q.

The purpose of this section is to prove the converse of this lemma.

Theorem 5.5 *Let $\Gamma \subset \operatorname{Aut} \Lambda$ be a Fuchsian group of type (g, k, m), $P = \Lambda / \Gamma$, $\Phi : \Lambda \to P$ be the natural projection, and*

$$v = \{a_i, \ b_i \ (i = 1, \dots, g), \ c_i \ (i = g + 1, \dots, n)\}$$

be a standard set of generators of the group $\pi_1(P, \Phi(q))$. Then $V = \Phi_q^{-1}(v)$ is a sequential set of type (g, k, m).

To prove this, we need some additional definitions and lemmas.

Let $\tilde{a}$ be a contour representing an element $a \in \pi_1(P, \Phi(q))$. We go along $\tilde{a}$ starting at $\Phi(q)$ and then "lift" this path to Λ starting at q. After making infinitely many circuits in both directions, we obtain a curve $\ell(\tilde{a}) \subset M$ with endpoints at the fixed points of the automorphism $A = \Phi_q^{-1}(a)$.

Lemma 5.4 *If $\tilde{a}$ has no self-intersections, then $h\ell(A)$ and $\ell(A)$ do not intersect each other for all $h \in \Gamma$.*

Proof Assume that $h\ell(A)$ and $\ell(A)$ intersect each other for $h \in \Gamma$. Then $h\ell(\tilde{a})$ and $\ell(\tilde{a})$ also intersect each other (see Fig. 5.10).

Fig. 5.10

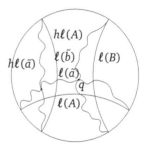

Lemma 5.5 *Let $\tilde{a}$, $\tilde{b}$ be contours representing elements $a, b \in \pi_1(P, \Phi(q))$. Assume that there exists a small deformation $\tilde{\tilde{a}}$, $\tilde{\tilde{b}}$ of these contours taking them to disjoint contours. Then*

$$\ell(\Phi_q^{-1}(a)) \cap \ell(\Phi_q^{-1}(b)) = \varnothing.$$

Proof If

$$\ell(\Phi_q^{-1}(a)) \cap \ell(\Phi_q^{-1}(b)) \neq \varnothing,$$

then $\ell(\tilde{a})$ and $\ell(\tilde{b})$ intersect each other in such a way that their intersection cannot be eliminated by a small deformation (see Fig. 5.9).

Lemma 5.6 *Let $\tilde{c}_1, \tilde{c}_2, \tilde{c}_3$ be contours without self-intersections that represent elements $c_1, c_2, c_3 \in \pi_1(P, \Phi(q))$ such that $c_1 \cdot c_2 \cdot c_3 = 1$. Assume that there exists a small deformation $\tilde{\tilde{c}}_1, \tilde{\tilde{c}}_2, \tilde{\tilde{c}}_3$ of these contours taking them to pairwise disjoint contours. Then either the set*

$$\{\Phi_q^{-1}(c_1), \Phi_q^{-1}(c_2), \Phi_q^{-1}(c_3)\}$$

or the set

$$\{\Phi_q^{-1}(c_3^{-1}), \Phi_q^{-1}(c_2^{-1}), \Phi_q^{-1}(c_1^{-1})\}$$

is sequential.

Proof Put $C_i = \Phi_q^{-1}(c_i)$. By Lemma 5.4, we have

$$\ell(C_1) \cap \ell(C_2) = \varnothing.$$

Assume that $\ell(C_1)$ and $\ell(C_2)$ are located as shown in Fig. 5.11.

Consider the arcs $\alpha = (\alpha_1, \alpha_2)$ and $\beta = (\beta_1, \beta_2)$ on $\partial \Lambda$. Then $C_1 \alpha \subset \alpha$, $C_2 \alpha \subset \alpha$, and hence $C_3^{-1} \alpha = C_1 C_2 \alpha \subset \alpha$. Similarly, $C_3^{-1} \beta \subset \beta$. Thus, $\ell(C_3)$ goes as the nonoriented curve in Fig. 5.11.

Fig. 5.11

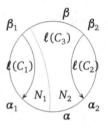

Fig. 5.12

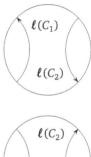

Fig. 5.13

Then, by Lemma 5.3, we have $C_1\ell(C_3) \subset N_1$ and, therefore, $C_1\beta_2 \in N_1$. This cannot be the case, since

$$C_1\beta_2 = C_3^{-1}C_2^{-1}\beta_2 = C_3^{-1}\beta \subset N_2.$$

Thus, $\ell(C_1)$ and $\ell(C_2)$ are located as shown in one of the Figs. 5.12 and 5.13.

A similar claim holds also for the pairs (C_2, C_3) and (C_3, C_1). It follows that either the set $\{C_1, C_2, C_3\}$ or the set $\{C_3^{-1}, C_2^{-1}, C_1^{-1}\}$ is sequential.

Proof of Theorem 5.5 Put

$$A_i = \Phi_q^{-1}(a_i), \quad B_i = \Phi_q^{-1}(b_i), \quad C_i = \Phi_q^{-1}(c_i).$$

Consider the sets

$$\{x_1, \ldots, x_{g+n}\} = \left\{a_1, b_1 a_1^{-1} b_1^{-1}, a_2, \ldots b_g a_g^{-1} b_g^{-1}, c_{g+1}, \ldots, c_n\right\}$$

and

$$\{X_1, \ldots, X_{g+n}\} = \left\{A_1, B_1 A_1^{-1} B_1^{-1}, A_2, \ldots, B_g A_g^{-1} B_g^{-1}, C_{g+1}, \ldots, C_n\right\}.$$

Applying Lemma 5.5 to the sets

$$\{x_1 \ldots x_{\ell-1}, x_\ell, x_{\ell+1} \ldots x_{g+n}\},$$

we see that either all sets of the form

$$\{X_1 \ldots X_{\ell-1}, X_\ell, X_{\ell+1} \ldots X_{g+n}\}$$

or all sets of the form

$$\{X_{g+n}^{-1} \ldots X_{\ell+1}^{-1}, X_{\ell}^{-1}, X_{\ell-1}^{-1} \ldots X_1^{-1}\}$$

are sequential, i.e., either $\{X_1, \ldots, X_{g+n}\}$ or $\{X_{g+n}^{-1}, \ldots, X_1^{-1}\}$ is a sequential set. However, according to Lemma 5.3, only in the first case the contours representing a_i, b_i, c_i are located in a neighborhood of p as shown in Fig. 5.8. Therefore, it is $\{X_1, \ldots, X_{g+n}\}$ that is a sequential set of type $(0, 2g + k, m)$, and hence

$$\{A_i, \ B_i \ (i = 1, \ldots, g), \ C_j \ (j = g + 1, \ldots, n\}$$

is a sequential set of type (g, k, m).

Exercise 5.19 Using Lemma 5.3 and Theorem 5.5, show that a set of automorphisms $\{C_1, \ldots, C_n\}$ is sequential if and only if the sets $\{C_1, \ldots, C_{n-2}, C\}$ and $\{C, C_{n-1}, C_n\}$ are sequential, where $C = (C_{n-1}, C_n)^{-1}$. State and prove a similar claim for arbitrary sequential sets of type (g, k, m).

5.9   Sequential Sets of Types (0, 3, 0), (0, 2, 1), and (0, 1, 2)

By Theorems 5.4 and 5.5, every sequential set of type (g, k, m) generates a Fuchsian group of type (g, k, m), and every Fuchsian group of type (g, k, m) is generated by a sequential set of type (g, k, m). In this section, we will find all conjugacy classes of sequential sets of types $(0, 3, 0)$, $(0, 2, 1)$, $(0, 1, 2)$. It is convenient to work with the upper half-plane.

Lemma 5.7 *Let*

$$C_1(z) = \lambda_1 z \quad (\lambda_1 > 1),$$

$$C_2(z) = \frac{(\lambda_2 \alpha - \beta)z + (1 - \lambda_2)\alpha\beta}{(\lambda_2 - 1)z + (\alpha - \lambda_2\beta)} \quad (\lambda_2 > 1),$$

$$C_3 = (C_1 C_2)^{-1}.$$

Then $\{C_1, C_2, C_3\}$ *is a sequential set if and only if*

$$0 < \left(\frac{\sqrt{\lambda_1} + \sqrt{\lambda_2}}{1 + \sqrt{\lambda_1 \lambda_2}} \right)^2 \beta \leq \alpha < \beta < \infty. \tag{5.1}$$

Moreover, C_3 *is a parabolic automorphism if and only if*

$$\left(\frac{\sqrt{\lambda_1} + \sqrt{\lambda_2}}{1 + \sqrt{\lambda_1 \lambda_2}} \right)^2 \beta = \alpha.$$

Proof By assumption,

$$C_3^{-1}(z) = C_1 C_2(z) = \lambda_1 \frac{(\lambda_2 \alpha - \beta) z + (1 - \lambda_2) \alpha \beta}{(\lambda_2 - 1) z + (\alpha - \lambda_2 \beta)}.$$

The fixed points of the automorphism C_3 are the roots of the equation $C_3^{-1}(x) = x$, i.e.,

$$(\lambda_2 - 1) x^2 - (\lambda_2 \beta - \alpha - \lambda_1 \beta + \lambda_1 \lambda_2 \alpha) x + \lambda_1 (\lambda_2 - 1) \alpha \beta = 0. \tag{5.2}$$

Therefore, C_3 is a hyperbolic or parabolic automorphism if and only if

$$(\lambda_2 \beta - \alpha - \lambda_1 \beta + \lambda_1 \lambda_2 \alpha)^2 - 4 \lambda_1 (\lambda_2 - 1)^2 \alpha \beta \geq 0,$$

with equality holding exactly for parabolic automorphisms.

Exercise 5.20 Show that the last inequality is equivalent to the inequality

$$(\alpha + \lambda_1 \lambda_2 \alpha - \lambda_1 \beta - \lambda_2 \beta)^2 - 4 \lambda_1 \lambda_2 (\beta - \alpha)^2 \geq 0,$$

which, in turn, holds only for

$$\alpha \geq \left(\frac{\sqrt{\lambda_1} + \sqrt{\lambda_2}}{1 + \sqrt{\lambda_1 \lambda_2}} \right)^2 \beta \quad \text{or} \quad \alpha \leq \left(\frac{\sqrt{\lambda_1} - \sqrt{\lambda_2}}{\sqrt{\lambda_1 \lambda_2} - 1} \right)^2 \beta.$$

Now let $\{C_1, C_2, C_3\}$ be a sequential set and $\bar{\alpha} \leq \bar{\beta}$ be the roots of (5.2). Then $0 < \alpha < \beta < \bar{\alpha}$ (see Fig. 5.14), and, by Vieta's formulas,

$$\frac{\lambda_2 \beta - \alpha - \lambda_1 \beta + \lambda_1 \lambda_2 \alpha}{2(\lambda_2 - 1)} = \frac{\bar{\alpha} + \bar{\beta}}{2} > \beta,$$

whence $\alpha(\lambda_1 \lambda_2 - 1) > \beta(\lambda_1 + \lambda_2 - 2)$. Besides, $\lambda_1 \lambda_2 - 1 > \lambda_1 + \lambda_2 - 2$, since $\lambda_i > 1$. Therefore,

$$\alpha > \frac{\lambda_1 + \lambda_2 - 2}{\lambda_1 \lambda_2 - 1} \beta > \frac{\lambda_1 + \lambda_2 - 2 + 2(1 - \sqrt{\lambda_1 \lambda_2})}{\lambda_1 \lambda_2 - 1 + 2(1 - \sqrt{\lambda_1 \lambda_2})} \beta$$

$$= \frac{\lambda_1 + \lambda_2 - 2\sqrt{\lambda_1 \lambda_2}}{\lambda_1 \lambda_2 - 2\sqrt{\lambda_1 \lambda_2} + 1} \beta = \left(\frac{\sqrt{\lambda_1} - \sqrt{\lambda_2}}{\sqrt{\lambda_1 \lambda_2} - 1} \right)^2 \beta.$$

Fig. 5.14

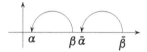

Thus,

$$\alpha \geq \left(\frac{\sqrt{\lambda_1} + \sqrt{\lambda_2}}{\sqrt{\lambda_1 \lambda_2} + 1} \right)^2 \beta.$$

Now let us prove the converse. Let

$$0 < \left(\frac{\sqrt{\lambda_1} + \sqrt{\lambda_2}}{\sqrt{\lambda_1 \lambda_2} + 1} \right)^2 \beta \leq \alpha < \beta < \infty,$$

and let $\bar{\alpha} \leq \bar{\beta}$ be the roots of (5.2). Then the inequality $\beta < \bar{\alpha}$ is equivalent to the pair of inequalities

$$\bar{\alpha} \cdot \bar{\beta} > \beta^2, \tag{5.3}$$

$$(\lambda_2 - 1)\beta^2 - (\lambda_2\beta - \alpha - \lambda_1\beta + \lambda_1\lambda_2\alpha)\beta + \lambda_1(\lambda_2 - 1)\alpha\beta > 0. \tag{5.4}$$

Inequality (5.3) is obvious by Vieta's formulas:

$$\bar{\alpha} \cdot \bar{\beta} = \lambda_1 \alpha \beta \geq \lambda_1 \left(\frac{\sqrt{\lambda_1} + \sqrt{\lambda_2}}{\sqrt{\lambda_1 \lambda_2} + 1} \right)^2 \beta^2 = \left(\frac{\lambda_1 + \sqrt{\lambda_1 \lambda_2}}{1 + \sqrt{\lambda_1 \lambda_2}} \right)^2 \beta^2 > \beta^2.$$

Inequality (5.4) follows from the fact that

$$\lambda_2\beta^2 - \beta^2 - \lambda_2\beta^2 + \alpha\beta + \lambda_1\beta^2 - \lambda_1\lambda_2\alpha\beta + \lambda_1\lambda_2\alpha\beta - \lambda_1\alpha\beta$$
$$= (\lambda_1 - 1)(\beta^2 - \alpha\beta) > 0.$$

Thus, $\beta < \bar{\alpha}$. Taking the limit as $\lambda_1 \to 1$, we see that $\bar{\alpha}$ is an attracting fixed point. Therefore, $\{C_1, C_2, C_3\}$ is a sequential set.

Exercise 5.21 Let

$$C_1(z) = \lambda z \ (\lambda > 1), \quad C_2(z) = \frac{(1 - a\gamma)z + a^2\gamma}{-\gamma z + (1 + a\gamma)} \ (\gamma > 0), \quad C_3 = (C_1 C_2)^{-1}.$$

Prove that $\{C_1, C_2, C_3\}$ is a sequential set if and only if $a\gamma \leq \frac{\sqrt{\lambda}+1}{\sqrt{\lambda}-1}$, with C_3 being parabolic if and only if $a\gamma = \frac{\sqrt{\lambda}+1}{\sqrt{\lambda}-1}$.

5.10   Sequential Sets of Type (1, 1, 0)

Lemma 5.8 *Let*

$$A(z) = \frac{(\lambda_A \alpha_A - \beta_A)z + (1 - \lambda_A)\alpha_A \beta_A}{(\lambda_A - 1)z + (\alpha_A - \lambda_A \beta_A)},$$

$$B(z) = \frac{(\lambda_B \alpha_B - \beta_B)z + (1 - \lambda_B)\alpha_B \beta_B}{(\lambda_B - 1)z + (\alpha_B - \lambda_B \beta_B)},$$

and

$$C^{-1} = [A, B](z) = \lambda z \quad (\lambda_A, \lambda_B, \lambda > 1).$$

Then $\{A, B, C\}$ is a sequential set of type $(1, 1, 0)$ if and only if

$$-\infty < \alpha_A < \beta_B < \beta_A < \alpha_B < 0, \tag{5.5}$$

$$\frac{\alpha_A}{\beta_A} < \sqrt{\lambda}, \quad \frac{\beta_B}{\alpha_B} < \sqrt{\lambda}, \tag{5.6}$$

$$\lambda_A = \frac{\alpha_A \sqrt{\lambda} - \beta_A}{\beta_A \sqrt{\lambda} - \alpha_A}, \quad \lambda_B = \frac{\beta_B \sqrt{\lambda} - \alpha_B}{\alpha_B \sqrt{\lambda} - \beta_B}, \tag{5.7}$$

$$\alpha_B \beta_B \lambda - [(\alpha_A + \beta_A)(\alpha_B + \beta_B) - \alpha_A \beta_A - \alpha_B \beta_B]\sqrt{\lambda} + \alpha_A \beta_A = 0, \tag{5.8}$$

and in this case

$$A(z) = \frac{(\alpha_A + \beta_A)\sqrt{\lambda}z - \alpha_A \beta_A(\sqrt{\lambda} + 1)}{(\sqrt{\lambda} + 1)z - (\alpha_A + \beta_A)},$$

$$B(z) = \frac{(\alpha_B + \beta_B)z - \alpha_B \beta_B(\sqrt{\lambda} + 1)}{(\sqrt{\lambda} + 1)z - (\alpha_B + \beta_B)\sqrt{\lambda}}.$$

Proof Let $\{A, B, C\}$ be a sequential set of type $(1, 1, 0)$. Then $\{A, BA^{-1}B^{-1}, C\}$ is a sequential set, and hence $-\infty < \alpha_A < \beta_B < \alpha_B < 0$ (see Fig. 5.15).

Fig. 5.15

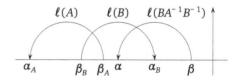

Consider the automorphism $\tilde{A} = (AB)A(AB)^{-1} = AC$. Let $\tilde{\alpha} < \tilde{\beta}$ be the fixed points of $\tilde{A}$. Then

$$\frac{(\lambda_A \alpha_A - \beta_A)\frac{1}{\lambda}z + (1 - \lambda_A)\alpha_A\beta_A}{(\lambda_A - 1)\frac{1}{\lambda}z + (\alpha_A - \lambda_A\beta_A)} = \tilde{A}(z) = \frac{(\lambda_A\tilde{\alpha} - \tilde{\beta})z + (1 - \lambda_A)\tilde{\alpha}\tilde{\beta}}{(\lambda_A - 1)z + (\tilde{\alpha} - \lambda_A\tilde{\beta})},$$

whence

$$\lambda_A\alpha_A - \beta_A = \lambda_A\tilde{\alpha} - \tilde{\beta}, \quad \lambda(1 - \lambda_A)\alpha_A\beta_A = (1 - \lambda_A)\tilde{\alpha}\tilde{\beta},$$

$$\lambda(\alpha_A - \lambda_A\beta_A) = \tilde{\alpha} - \lambda_A\tilde{\beta}.$$

We obtain

$$\lambda_A = \frac{\alpha_A\sqrt{\lambda} - \beta_A}{\beta_A\sqrt{\lambda} - \alpha_A}, \quad \lambda_B = \frac{\beta_B\sqrt{\lambda} - \alpha_B}{\alpha_B\sqrt{\lambda} - \beta_B}.$$

In particular,

$$\sqrt{\lambda} = \frac{\beta_A - \alpha_A\lambda_A}{\alpha_A - \beta_A\lambda_A} > \frac{-\alpha_A\lambda_A}{-\beta_A\lambda_A} = \frac{\alpha_A}{\beta_A}, \quad \text{and similarly} \quad \sqrt{\lambda} > \frac{\beta_B}{\alpha_B}.$$

Substituting the obtained values for λ_A, λ_B, we have

$$A(z) = \frac{(\alpha_A + \beta_A)\sqrt{\lambda}z - \alpha_A\beta_A(\sqrt{\lambda} + 1)}{(\sqrt{\lambda} + 1)z - (\alpha_A + \beta_A)},$$

$$B(z) = \frac{(\alpha_B + \beta_B)z - \alpha_B\beta_B(\sqrt{\lambda} + 1)}{(\sqrt{\lambda} + 1)z - (\alpha_B + \beta_B)\sqrt{\lambda}}.$$

The relation $[A, B](z) = \lambda z$ implies (5.8).

Now assume that conditions (5.5)–(5.8) are satisfied. Put

$$\tilde{A}(z) = \frac{(\alpha_A + \beta_A)\sqrt{\lambda}z - \alpha_A\beta_A(\sqrt{\lambda} + 1)}{(\sqrt{\lambda} + 1)z - (\alpha_A + \beta_A)},$$

$$\tilde{B}(z) = \frac{(\alpha_B + \beta_B)z - \alpha_B\beta_B(\sqrt{\lambda} + 1)}{(\sqrt{\lambda} + 1)z - (\alpha_B + \beta_B)\sqrt{\lambda}}.$$

Then $\tilde{A}(\alpha_A) = \alpha_A$, $\tilde{A}(\beta_A) = \beta_A$, $\tilde{B}(\alpha_B) = \alpha_B$, $\tilde{B}(\beta_B) = \beta_B$. By (5.8), we have $[\tilde{A}, \tilde{B}] = C^{-1}$. Hence $\{\tilde{A}, \tilde{B}\tilde{A}^{-1}\tilde{B}^{-1}, C\}$ is a sequential set. Thus, $\{\tilde{A}, \tilde{B}, C\}$ is a sequential set of type $(1, 1, 0)$. Since the fixed points coincide, it follows that

$$\tilde{A}z = \frac{(\tilde{\lambda}_A\alpha_A - \beta_A)z + (1 - \tilde{\lambda}_A)\alpha_A\beta_A}{(\tilde{\lambda}_A - 1)z + (\alpha_A - \lambda_A\beta_A)};$$

besides, as we have already proved,

$$\tilde{\lambda}_A = \frac{\alpha_A\sqrt{\lambda} - \beta_A}{\beta_A\sqrt{\lambda} - \alpha_A} = \lambda_A.$$

Therefore, $\tilde{A} = A$. In a similar way, $\tilde{B} = B$.

5.11 Fricke–Klein–Teichmüller Type Spaces

To describe the moduli spaces $M_{g,k,m}$, it is convenient to use an auxiliary space $T_{g,k,m}$ homeomorphic to $\mathbb{R}^{6g+3k+2m-6}$. It was first constructed in the technically difficult two-volume monograph [5] by Fricke and Klein in terms of lengths of geodesics of hyperbolic metrics on Riemann surfaces. Another description of this space was given by [27] in terms of the theory of quasi-conformal maps, which he developed as a generalization of the theory of holomorphic univalent maps.

We will assume that $T_{g,k,m}$ is the set of conjugacy classes of sequential sets of type (g, k, m) parametrized by attracting fixed points $\alpha \in \mathbb{R}$, repelling fixed points $\beta \in \mathbb{R}$, and shift parameters $\lambda > 1$.

Theorem 5.6 *The space $T_{g,k,m}$ is isomorphic to $\mathbb{R}^{6g+3k+2m-6}$ as a real manifold.*

Proof In a conjugacy class of sequential sets of type $(0, 3, 0)$ there is exactly one set (C_1, C_2, C_3) such that $C_1(z) = \lambda_1 z$, $\lambda_1 > 1$, and $\beta_{C_2} = 1$. By Lemma 5.7, the set of sequential sets (C_1, C_2, C_3) satisfying this condition is determined by numbers $(\lambda_1, \lambda_2, \alpha)$ where $\lambda_1, \lambda_2 > 1$ and $\left(\frac{\sqrt{\lambda_1}+\sqrt{\lambda_2}}{1+\sqrt{\lambda_1\lambda_2}}\right)^2 < \alpha < 1$. Thus, $T_{0,3,0}$ is homeomorphic to $\mathbb{R}^3$.

Now let us prove by induction that the space $T_{0,k,0}$ is homeomorphic to $\mathbb{R}^{3k-6}$. By Exercise 5.19, the space $T_{0,k,0}$ coincides with the space of conjugacy classes of pairs of sequential sets $\{C_1, \ldots, C_{k-2}, C^{-1}\}$ and $\{C, C_{k-1}, C_k\}$. In such a conjugacy class there is exactly one pair for which $C(z) = \lambda z$, $\lambda > 1$, and $\beta_{C_{k-2}} = -1$. By Lemma 5.7, the set of sequential sets (C, C_{k-1}, C_k) is determined by positive parameters $(\alpha_{C_{k-1}}, \beta_{C_{k-1}}, \lambda_{C_{k-1}})$ satisfying the constraints

$$\left(\frac{\sqrt{\lambda_C}+\sqrt{\lambda_{C_{k-1}}}}{1+\sqrt{\lambda_C\lambda_{C_{k-1}}}}\right)^2 \beta_{C_{k-1}} < \alpha_{C_{k-1}} < \beta_{C_{k-1}}, \quad \lambda_{C_{k-1}} > 1.$$

These constraints define a domain homeomorphic to $\mathbb{R}^3$. Now, using the induction hypothesis, we see that the set $T_{0,k,0}$ is homeomorphic to $T_{0,k-1,0} \times \mathbb{R}^3 = \mathbb{R}^{3k-6}$.

In a conjugacy class of sequential sets of type $(1, 1, 0)$ there is exactly one set (A, B, C) for which $C(z) = \lambda z$, $\lambda > 1$, and $\alpha_B = -1$. By Lemma 5.8, the set of sequential sets (A, B, C) satisfying these conditions is determined by numbers $(\lambda, \alpha_A, \beta_A, \beta_B)$ where $-\infty < \alpha_A < \beta_B < \beta_A < -1$, $\frac{\alpha_A}{\beta_A} < \sqrt{\lambda}$, $\frac{\beta_B}{-1} < \sqrt{\lambda}$, and

$$-\beta_B \lambda - [(\alpha_A + \beta_A)(-1 + \beta_B) - \alpha_A \beta_A + \beta_B]\sqrt{\lambda} + \alpha_A \beta_A = 0.$$

Thus, $T_{1,1,0}$ is homeomorphic to $\mathbb{R}^3$.

Now we prove the theorem for surfaces of arbitrary type $(g, k, 0)$. By Exercise 5.19, a point of the space $T_{g,k,0}$ consists of the conjugacy class of a sequential set $\{C_1, \ldots, C_{g+k}\}$ of type $(0, g + k, 0)$ and the conjugacy classes of sequential sets $(A_1, B_1, \tilde{C}_1), \ldots, (A_g, B_g, \tilde{C}_g)$ of type $(1, 1, 0)$ such that the hyperbolic shifts C_i and $\tilde{C}_i$ are conjugate for all i. By Lemma 5.8, for the automorphism $C: z \mapsto \lambda z$, the set of sequential sets (A, B, C) of type $(1, 1, 0)$ is determined by numbers $(\alpha_A, \beta_A, \alpha_B, \beta_B)$ where

$$-\infty < \alpha_A < \beta_B < \beta_A < \alpha_B < 0, \quad \frac{\alpha_A}{\beta_A} < \sqrt{\lambda}, \quad \frac{\beta_B}{\alpha_B} < \sqrt{\lambda},$$

and

$$\alpha_B \beta_B \lambda - [(\alpha_A + \beta_A)(\alpha_B + \beta_B) - \alpha_A \beta_A - \alpha_B \beta_B]\sqrt{\lambda} + \alpha_A \beta_A = 0.$$

Thus, $T_{g,k,0}$ is homeomorphic to the space

$$T_{0,g+k,0} \times (\mathbb{R}^3)^g \cong \mathbb{R}^{6g+3k-6}.$$

The general case of the space $T_{g,k,m}$ can be treated in a similar way and is left to the reader as an exercise.

5.12   The Moduli Space $M_{g,k,m}$

Consider the space $\tilde{T}_{g,k,m}$ of all sequential sets of type (g, k, m). A sequential set $\{A_1, B_1, \ldots, A_g, B_g, C_1, \ldots, C_n\} \in \tilde{T}_{g,k,m}$ generates a group Γ and, by Theorem 5.4, a Riemann surface $\Lambda/\Gamma \in M_{g,k,m}$. The obtained map $\tilde{\Phi}: \tilde{T}_{g,k,m} \to M_{g,k,m}$ is surjective by Theorem 5.5.

The group $\mathrm{Aut}\, \Lambda$ acts on $\tilde{T}_{g,k,m}$ by sending a set

$$\{A_1, B_1, \ldots, A_g, B_g, C_1, \ldots, C_n\}$$

to the conjugate set

$$h\{A_1, B_1, \ldots, A_g, B_g, C_1, \ldots, C_n\}h^{-1}, \quad h \in \mathrm{Aut}\, \Lambda.$$

By definition, the quotient space $\tilde{T}_{g,k,m}/\operatorname{Aut}\Lambda$ coincides with $T_{g,k,m}$. To conjugate sequential sets there correspond isomorphic Riemann surfaces. Thus, the map $\tilde{\Phi}$ generates a surjective map $\Phi: T_{g,k,m} \to M_{g,k,m}$.

Let us find out which points of the space $T_{g,k,m}$ go to the same Riemann surface $P = \Lambda/\Gamma$. Let Hom P be the group of autohomeomorphisms of the surface P and IHom $P \subset$ Hom P be the subgroup of autohomeomorphisms isotopic to the identity. The quotient Mod $P = $ Hom $P/$IHom P is called the *mapping class group*. It plays an important role in low-dimensional topology. For example, in terms of this group one can give a classification of three-dimensional topological manifolds.

Now let $P = \Lambda/\Gamma$ be a Riemann surface of type (g, k, m) and $T(P)$ be a set of sequential sets of type (g, k, m) that generates the Fuchsian group obtained from Γ by conjugation by an element of $\operatorname{Aut}(H)$. Standard bases of the groups $\pi_1(P, q)$ and $\pi_1(P, q')$ are said to be *equivalent* if the first basis turns into the second one as we continuously change the point q. Denote by $t(P)$ the set of equivalence classes of standard bases of the fundamental group of P. Lemma 5.3 and Theorem 5.5 establish a natural one-to-one correspondence between the sets $T(P)$ and $t(P)$. The mapping class group Mod P acts transitively on the set $t(P)$, and hence on the set $T(P)$.

Let

$$P = \Lambda/\Gamma, \quad V \in T(P), \quad P' = \Lambda/\Gamma', \quad V' \in T(P'), \quad \text{and} \quad \varphi \in \operatorname{Mod} P.$$

An element $D \in \Gamma$ can be represented as a product of elements of the sequential set V. Replacing the factors in this product by the corresponding elements of the sequential set V', we obtain an element $D' \in \Gamma'$. The element $\varphi(D)$, too, can be represented as a product of elements of the sequential set V. Replacing the factors in this product by the corresponding elements of the sequential set V', we obtain an element which we denote by $\varphi(D')$. The correspondence $D' \mapsto \varphi(D')$ defines an action of the group Mod P on $T(P')$ and a natural isomorphism between the groups Mod P and Mod P'. Denote by $\operatorname{Mod}_{g,k,m}$ the group obtained by identifying the groups of the form Mod P via such isomorphisms.

Then we obtain the following result.

Lemma 5.9 *There is a natural action of the group* $\operatorname{Mod}_{g,k,m}$ *on the space* $T_{g,k,m}$ *which turns a map* $\Phi: T_{g,k,m} \to M_{g,k,m}$ *into a one-to-one map*

$$T_{g,k,m}/\operatorname{Mod}_{g,k,m} \to M_{g,k,m}.$$

Lemma 5.10 *The group* $\operatorname{Mod}_{g,k,m}$ *acts on* $T_{g,k,m}$ *discretely by smooth maps.*

Proof According to our definitions, an element from $\operatorname{Mod}_{g,k,m}$ is determined by a representation of elements of one sequential set of generators as products of elements of another sequential set of generators. Therefore, the parameters determining the new sequential set can be expressed in terms of the parameters determining the old sequential set by analytic formulas.

To prove that the action of the group $\mathrm{Mod}_{g,k,m}$ on the set $T(\Lambda/\Gamma)$ is discrete, consider the set $L(\Gamma)$ of shift parameters of all transformations from Γ. Using Lobachevsky geometry, one can easily show that the set $L(\Gamma)$ is discrete, uniquely determines the Riemann surface Λ/Γ, and does not depend on the sequential set generating the group Γ, but can be computed in terms of its parameters. Moreover, a small change in the parameters determining the sequential set results in a small change in $L(\Gamma)$. Thus, a small neighborhood of a point from T contains no other points of the orbit of this point under the action of the group $\mathrm{Mod}_{g,k,m}$.

Theorem 5.7 *The moduli space $M_{g,k,m}$ has a natural structure of a connected real-analytic space of the form $\mathbb{R}^{6g+3k+2m-6}/\mathrm{Mod}_{g,k,m}$ where $\mathrm{Mod}_{g,k,m}$ is a discrete group of real-analytic maps.*

Proof By Lemmas 5.9 and 5.10, the set $M_{g,k,m}$ can be naturally identified with the real-analytic space $T_{g,k,m}/\mathrm{Mod}_{g,k,m}$. Besides, by Theorem 5.6, the space $T_{g,k,m}$ is isomorphic to $\mathbb{R}^{6g+3k+2m-6}$.

Remark 5.1 One can prove that the moduli space $M_{g,0,m}$ has even a natural complex-analytic structure. Singularities of the space $M_{g,k,m}$ correspond to Riemann surfaces that have nontrivial holomorphic automorphisms.

Chapter 6
Compact Riemann Surfaces

6.1 The Riemann–Hurwitz Formula

In this section, we will consider only compact Riemann surfaces of genus g, i.e., surfaces of type $(g, 0, 0)$. Such a surface is homeomorphic to a sphere with g holes in which every boundary contour is glued to the boundary contour of a torus with a hole. Recall that a complex structure on a surface is defined by a holomorphic atlas of local charts. A map between surfaces is said to be holomorphic if it is holomorphic in every local chart.

Example 6.1 The Riemann sphere $\bar{\mathbb{C}} = \mathbb{C} \bigcup \infty$ with the atlas $\{(U_1, z_1), (U_2, z_2)\}$ where $U_1 = \{z \in \mathbb{C} \mid |z| < 2,\ z_1(z) = z\}$ and $U_2 = \{z \in \mathbb{C} \mid |z| > \frac{1}{2},\ z_2(z) = \frac{1}{z}\}$.

Example 6.2 The complex torus $T = \mathbb{C}/\Gamma$ where Γ is the group of translations generated by the shifts $z \mapsto z + 1$ and $z \mapsto z + \tau$ with $\Im \tau > 0$.

Example 6.3 The surface

$$P = \{(x, y) \in \bar{\mathbb{C}}^2 \mid y^2 = f(x, y)\}$$

where $f(x, y) = x^{2m+1} + a_{2m}x^{2m} + \ldots + a_1 x + a_0$. A holomorphic atlas consists of the charts (U_v, z_v) where U_v is a small neighborhood of $v = (x, y)$ and $z_v(x, y) = x$ for $x \in \mathbb{C}$ such that $f(x) \neq 0$, $z_v(x, y) = y$ for $x \in \mathbb{C}$ such that $f(x) = 0$, and $z_v(x, y) = \frac{1}{y}$ for $x = \infty$. Such surfaces are called *hyperelliptic*.

The properties of a holomorphic map $f : P \to Q$ in a neighborhood of a point $p \in P$ are described by the function

$$f_{\alpha\beta} = w_\beta f z_\alpha^{-1} : z_\alpha \left(U_\alpha \bigcap f^{-1}(V_\beta) \right) \to \mathbb{C},$$

© Springer Nature Switzerland AG 2019
S. M. Natanzon, *Complex Analysis, Riemann Surfaces and Integrable Systems*,
Moscow Lectures 3, https://doi.org/10.1007/978-3-030-34640-9_6

where (U_α, z_α) and (V_β, w_β) are local charts containing the points p and q, respectively. Like any holomorphic function, it can be represented as a series $f_{\alpha\beta} = \sum_{k=n}^{\infty} a_k z^k$ where $a_n \neq 0$. The number n is called the *ramification degree of the map f at the point p* and denoted by $\deg_p f$.

Exercise 6.1 Show that the ramification degree n of a holomorphic map f at a point p does not depend on the choice of local charts (U_α, z_α) and (V_β, w_β), and that $f_{\alpha\beta}(z_\alpha) = z^n$.

A point $p \in P$ is called a *critical point*, or *ramification point*, of a holomorphic map f if $\deg_p f > 1$, i.e., $f'_{\alpha\beta}(z_\alpha(p)) = 0$.

A point $q \in Q$ is called a *critical value* of a holomorphic map f if the preimage $f^{-1}(q)$ has at least one critical point.

Exercise 6.2 Show that the number of critical points of a holomorphic map $f: P \to Q$ between (compact) Riemann surfaces is finite. Let $V \subset Q$ be an open domain whose closure is connected, simply connected, and contains no critical values of f. Show that the preimage $f^{-1}(V)$ breaks into finitely many connected components with f being a homeomorphism on each of them.

Lemma 6.1 (Lemma–Definition) *Let $f: P \to Q$ be a holomorphic map between Riemann surfaces. Then the number of preimages $|f^{-1}(q)|$ is the same for all noncritical values $q \in Q$. This number is called the degree of f and denoted by $\deg f$.*

Proof Let $q_1, q_2 \in Q$ be noncritical values and consider a connected, simply connected, open domain V containing them whose closure contains no critical values of f. Then, by Exercise 6.2, each connected component of the preimage $f^{-1}(V)$ contains exactly one preimage from $f^{-1}(q_i)$. Therefore, the number of preimages $|f^{-1}(q_i)|$ coincides with the number of connected components of the preimage $f^{-1}(V)$.

Exercise 6.3 Consider a holomorphic map $f: P \to Q$ between compact Riemann surfaces. Show that $\deg f = \sum_{p \in f^{-1}(q)} \deg_p f$ for every point $q \in Q$.

Theorem 6.1 (Riemann–Hurwitz Formula) *Let $f: P \to Q$ be a holomorphic map from a Riemann surface P of genus $\tilde{g}$ to a Riemann surface Q of genus g. Then*

$$\tilde{g} = (\deg f)(g - 1) + \frac{1}{2} \sum_{p \in P} (\deg_p f - 1) + 1.$$

Proof Consider a triangulation of the surface Q whose vertices include all critical values of f. Assume that it has F faces, E edges, and V vertices. The preimage of this triangulation is a triangulation of the surface P. It consists of $(\deg f) F$ triangles and $(\deg f) E$ edges. The number of vertices in this triangulation is equal to $(\deg f) V - \sum_{p \in P} (\deg_p f - 1)$, since every triangle with vertex $q \in Q$ produces exactly $\deg_p f$ triangles having a vertex at $p \in f^{-1}(q)$. Calculating the Euler characteristics yields $2 - 2g = F - E + V$ and

$$2 - 2\tilde{g} = (\deg f) F - (\deg f) E + (\deg f) V - \sum_{p \in P} (\deg_p f - 1),$$

whence $\deg f (2 - 2g) - (2 - 2\tilde{g}) = \sum_{p \in P} (\deg_p f - 1)$.

6.2 Meromorphic Functions and Differentials

Let P be a Riemann surface given by a holomorphic atlas of local charts

$$\{(U_\alpha, z_\alpha) \mid \alpha \in \mathscr{A}\}.$$

A holomorphic map $f \colon P \to \bar{\mathbb{C}}$ to the Riemann sphere $\bar{\mathbb{C}} = \mathbb{C} \cup \infty$ is called a *meromorphic function.*

This means that in a neighborhood of a point $p \in U_\alpha$ the map f has the form $f|_{U_\alpha}(p) = \sum_{i=k}^{\infty} a_i (z_\alpha(p))^i$ where $a_k \neq 0$. The point p is called a *zero of order k* (for $k > 0$) or a *pole of order $-k$* (for $k < 0$) of the function f. By Exercise 6.3, every meromorphic function has a pole and a zero. A pole of order 1 is said to be *simple.*

Exercise 6.4 Show that the projections $(x, y) \mapsto x$ and $(x, y) \mapsto y$ on a hyperelliptic Riemann surface P_F are meromorphic functions. Find the poles of these functions and their orders.

Exercise 6.5 Show that the set of meromorphic functions on a Riemann surface P is a field $\mathscr{M}(P)$.

A *meromorphic differential* on a Riemann surface P is defined by a family of meromorphic functions $\{f_\alpha \colon U_\alpha \to \mathbb{C} \mid \alpha \in \mathscr{A}\}$ on local charts $\{(U_\alpha, z_\alpha)\}$ such that $\frac{f_\alpha}{f_\beta} = (z_\beta z_\alpha^{-1})'_{z_\alpha}$ on $U_\alpha \cap U_\beta$. It is convenient to write the meromorphic differential ω corresponding to this family of functions as $f_\alpha \, dz_\alpha$. The meromorphic differentials form a vector space over the field of complex numbers.

Exercise 6.6 Show that the set of meromorphic differentials depends only on the Riemann surface. In other words, for an equivalent holomorphic atlas of local charts $\{(V_\beta, w_\beta) \mid \beta \in \mathscr{B}\}$ there exists a uniquely determined holomorphic differential $h_\alpha \, dw_\alpha$ such that $\frac{f_\alpha}{h_\beta} = (w_\beta z_\alpha^{-1})'_{z_\alpha}$ on $U_\alpha \cap V_\beta$.

Zeros and poles of the meromorphic functions f_α are called *zeros* and *poles* of the meromorphic differential $f_\alpha \, dz_\alpha$. The orders of zeros and poles of f_α are called the *orders* of zeros and poles of $f_\alpha\, dz_\alpha$. In other words, if $f_\alpha(p) = \sum\limits_{i=k}^{\infty} a_i(z_\alpha(p))^i$ with $a_k \neq 0$ and $k \neq 0$, then the point p is called a *zero of order* k (for $k > 0$) or a *pole of order* $-k$ (for $k < 0$) of the differential ω. The set of zeros and poles of a meromorphic differential and their orders do not change when we replace a holomorphic atlas of local charts by an equivalent atlas. A meromorphic differential that has no poles is called a *holomorphic differential*.

Theorem 6.2 *Let P be an arbitrary Riemann surface of genus g. Then*

(1) *on P there exist at least g linearly independent holomorphic differentials;*
(2) *for every point $p \in P$ and every integer $j > 1$ there exists a meromorphic differential that has a single pole at p and has the form $\left(\dfrac{1}{z^j} + \sum\limits_{s=0}^{\infty} a_s z^s\right) dz$ in some local chart;*
(3) *for every pair of points $p_1 \neq p_2 \in P$ there exists a meromorphic differential that has simple poles at p_1 and p_2 and no other poles.*

This remarkable, difficult, and very important theorem [26, Chap. 8] is, along with the uniformization theorem, one of the main achievements of the late nineteenth century mathematics. Contributions to its proof were made by Riemann, Weierstrass, Poincaré, Klein, Hilbert, H. Weyl, Koebe. We will not prove this theorem in this course, but we will heavily use it.

Exercise 6.7 Show that for every polynomial $g(x)$, the formula $\omega = \dfrac{g(x)\, dx}{y}$ describes a meromorphic differential on any hyperelliptic Riemann surface. Find the poles of this differential and their orders. Illustrate the last theorem by considering hyperelliptic Riemann surfaces.

6.3 Plane Algebraic Curves

Let $F(z, w)$ be a polynomial in two variables. The set

$$P_F = \{(z, w) \in \mathbb{C} \mid F(z, w) = 0\}$$

is called a *plane affine complex algebraic curve*. For a plane affine complex algebraic curve, a holomorphic analog of the implicit function theorem holds.

Theorem 6.3 *Let $F(z_0, w_0) = 0$ and $\dfrac{\partial F}{\partial w}(z_0, w_0) \neq 0$. Then in a neighborhood of z_0 there exists a unique univalent holomorphic function $w = w(z)$ such that $w_0 = w(z_0)$ and $F(z, w(z)) = 0$.*

Proof Let $z = x + iy$, $w = u + iv$, $F = f + ig$. If we regard the function $F(z_0, w)$ as a map $(u, v) \mapsto (f, g)$, then its Jacobian at the point $w_0 = (u_0, v_0)$ equals

$$\det \begin{pmatrix} \dfrac{\partial f}{\partial u} & \dfrac{\partial f}{\partial v} \\ \dfrac{\partial g}{\partial u} & \dfrac{\partial g}{\partial v} \end{pmatrix} = \det \begin{pmatrix} \dfrac{\partial f}{\partial u} & -\dfrac{\partial g}{\partial u} \\ \dfrac{\partial g}{\partial u} & \dfrac{\partial f}{\partial u} \end{pmatrix} = \left(\dfrac{\partial f}{\partial u} \right)^2 + \left(\dfrac{\partial g}{\partial u} \right)^2 = \left| \dfrac{\partial F}{\partial w} \right|^2$$

and, therefore, does not vanish. Hence, by the implicit function theorem, in a neighborhood of z_0 there exist functions

$$u(z) = u(x, y), \quad v(z) = v(x, y), \quad w(x, y) = u(x, y) + i v(x, y)$$

that generate a one-to-one map $(x, y) \mapsto (u, v)$ and satisfy

$$F(z, w(z, \bar{z})) = F(z, w(x, y)) = 0.$$

Besides,

$$0 = \frac{\partial F}{\partial \bar{z}} = \frac{\partial F}{\partial z} \frac{\partial z}{\partial \bar{z}} + \frac{\partial F}{\partial w} \frac{\partial w}{\partial \bar{z}} = \frac{\partial F}{\partial w} \frac{\partial w}{\partial \bar{z}}.$$

The inequality $\frac{\partial F}{\partial w} \neq 0$ implies that $\frac{\partial w}{\partial \bar{z}} = 0$.

Corollary 6.1 *In a neighborhood of a point (z_0, w_0) where $\frac{\partial F}{\partial w}(z_0, w_0) \neq 0$, the map $(z, w(z)) \mapsto z$ defines a local chart on P_F. The set of all such charts is a holomorphic atlas on the set $P_F \setminus \Sigma_w$ where $\Sigma_w = \{(z, w) \in P_F \mid \frac{\partial F}{\partial w}(z, w) = 0\}$.*

In a similar way, one defines a holomorphic atlas of local charts on $P_F \setminus \Sigma_z$ where $\Sigma_z = \left\{ (z, w) \in P_F \mid \frac{\partial F}{\partial z}(z, w) = 0 \right\}$.

Exercise 6.8 Show that these atlases are equivalent on $P_F \setminus \left(\Sigma_z \bigcup \Sigma_w \right)$.

A curve P_F is said to be *nonsingular* if $\left| \frac{\partial F}{\partial z}(z, w) \right| + \left| \frac{\partial F}{\partial w}(z, w) \right| > 0$ for all points $(z, w) \in P_F$. In this case, the atlases constructed above define a Riemann surface structure on P_F. However, the manifold P_F is not compact. In particular, a sequence $\{(z_n, w_n) \in \mathbb{C} \mid F(z_n, w_n) = 0\}$ where $z_n \to \infty$ does not converge to any point of P_F. To compactify P_F, consider the set $c_R = \{z \in \mathbb{C} \mid |z| > R\}$ where R is greater than the absolute value of every critical value of the function $h(z, w) = z$ on P_F. The set $h^{-1}(c_R)$ is an unramified covering of the cylinder c_R. Extend the covering map to a map $h^{-1}(c_R) \cup D \to c_R \cup \infty$ by adding one point to each connected component of the preimage $h^{-1}(c_R)$. This extension, together with the map h, generates a map $\tilde{h}: P \to \bar{\mathbb{C}}$ from the surface $\bar{P}_F = P_F \bigcup D$ to the Riemann sphere.

Exercise 6.9 Show that the surface $\bar{P}_F$ is compact. Find its genus for a generic polynomial F of degree d in the case where $\bar{P}_F$ is connected.

Exercise 6.10 Show that the surface $\bar{P}_F$ has a Riemann surface structure with respect to which $\tilde{h}$ is a holomorphic map.

The Riemann surface $\bar{P}_F$ is called the *Riemann surface of the polynomial* $F(z, w)$ *and the algebraic curve* $F(z, w) = 0$.

Exercise 6.11 Show that every rational function $R(z, w)$ generates a meromorphic function on $\bar{P}_F$.

As we have just seen, a plane affine algebraic curve can be regarded as a Riemann surface with a distinguished pair of meromorphic functions. These curves are objects of a category whose morphisms are rational changes of variables $z \mapsto \tilde{z}(z, w)$, $w \mapsto \tilde{w}(z, w)$. An isomorphism class in this category no longer depends on the pair of functions. Moreover, below we will prove that this category is isomorphic to the category of compact Riemann surfaces.

6.4 The Field of Algebraic Functions

Lemma 6.2 *Let* $f, h \colon P \to \bar{\mathbb{C}}$ *be meromorphic functions on a Riemann surface* P *and* $\deg h = n$. *Then there exist rational functions* $r_k \colon \bar{\mathbb{C}} \to \bar{\mathbb{C}}$ $(k = 1, \ldots, n)$ *such that* $f^n + r_1(h) f^{n-1} + \ldots + r_{n-1}(h) f + r_n(h) = 0$ *on* P.

Proof Consider the symmetric functions

$$\sigma_k(x_1, \ldots, x_n) = (-1)^k \sum_{1 \leq i_1 < i_2 < \ldots < i_k \leq n} x_{i_1} x_{i_2} \ldots x_{i_k}.$$

For every point $z \in \bar{\mathbb{C}}$, consider the preimage $h^{-1}(z) = \bigcup p_i(z)$ and put

$$r_k(z) = \sigma_k(f(p_1(z)), \ldots, f(p_n(z))).$$

(This function is well defined, since the value $\sigma_k(f(p_1(z)), \ldots, f(p_n(z)))$ does not depend on the ordering of the points of $h^{-1}(z)$.) Let

$$r_k(h)(p) = \sigma_k(f(p_1(h(p))), \ldots, f(p_n(h(p)))).$$

By Vieta's formulas,

$$(f^n + r_1(h) f^{n-1} + \ldots + r_{n-1}(h) f + r_n(h))(p)$$
$$= (f(p) - f(p_1(h(p))))(f(p) - f(p_2(h(p)))) \ldots (f(p) - f(p_n(h(p)))) = 0$$

for $p \in P$, since at least one of the bracketed expressions vanishes.

A polynomial $F(y, z)$ is said to be *reducible* if it can be written as a product $F(y, z) = F_1(y, z)F_2(y, z)$ where $F_1(y, z)F_2(y, z)$ are polynomials of positive degree with rational coefficients, $F_i(y, z) = f_i^{n_i}(y)z^{n_i} + \ldots + f_i^0(y)$. Otherwise, $F(y, z)$ is said to be *irreducible*.

Exercise 6.12 Show that the Riemann surface of a reducible polynomial is not connected.

Theorem 6.4 *Let* $h: P \to \bar{\mathbb{C}}$ *be a meromorphic function of degree* n *on a connected Riemann surface* P. *Then there exist a meromorphic function* $f: P \to \bar{\mathbb{C}}$ *and a nonsingular irreducible polynomial* $F(y, z)$ *such that* $F(f, h) = 0$ *on* P.

Proof Let $z_0 \in \bar{\mathbb{C}}$ be a noncritical value, i.e., $h^{-1}(z_0) = \bigcup_{i=1}^{n} p_i$ where $p_i \neq p_j$ for $i \neq j$. The function h generates a local chart z_i in a neighborhood of p_i such that $h(p) = (z_i(z) - z_0)$. Consider a meromorphic differential ω_i that is holomorphic outside p_i and such that

$$\omega_i = \left(\frac{1}{(z_i - z_0)^2} + \sum_{n=0}^{\infty} (z_i - z_0)^n \right) dz_i \quad \text{in a neighborhood of } p_i.$$

Consider the meromorphic function

$$f(p) = (h(p) - z_0)^2 \left(c_1 \frac{\omega_1}{dh} + \ldots + c_n \frac{\omega_n}{dh} \right) \quad \text{where } c_i \neq c_j.$$

Then the points p_i are not critical and $f(p_i) = c_i$. By the previous lemma, there exists a polynomial $F(y, z)$ such that $F(f, h) = 0$ on P. This polynomial is nonsingular for a generic collection of numbers $\{c_i\}$.

Let us prove that the polynomial $F(y, z)$ is irreducible. Assume that $F = F_1 F_2$ where $F_1(y, z)$ and $F_2(y, z)$ are polynomials of positive degree in z whose coefficients are rational functions in y. By construction, $F(f(p_1), h(p_1)) = 0$. Assume that $F_1(f(p_1), h(p_1)) = 0$. In a neighborhood of p_1, consider the local charts generated by the functions h and f. In a neighborhood of $c_1 = f(p_1)$, consider the holomorphic transition map $H = hf^{-1}$ between the local charts. Then $H(f(p)) = h(p)$ and $F_1(w, H(w)) = F_1(f(p), h(p)) = 0$ for $w = f(p)$.

Now consider a path Γ_i lying in P and connecting the points p_1 and p_i. Set $\gamma_i = f(\Gamma_i)$ and consider the analytic continuation of the function $H(u)$ along the path γ_i. Then we obtain a holomorphic function $\tilde{H}(\tilde{u})$ in a neighborhood of the point $c_i = f(p_i)$ such that $F_1(\tilde{u}, \tilde{H}(\tilde{u})) = 0$. Therefore, $F_1(c_i, z_0) = F_1(f(p_i), h(p_i)) = 0$ and $\deg_z F_1 = n$, whence $\deg_z F_2 = 0$.

Corollary 6.2 *Every compact connected Riemann surface is the Riemann surface of some plane affine nonsingular irreducible complex algebraic curve.*

Exercise 6.13 Assume that a nonsingular polynomial $F(y, z)$ cannot be written as a product of two polynomials of positive degree in z whose coefficients are rational functions in y. Then it cannot be written as a product of two polynomials of positive degree in y whose coefficients are rational functions in z.

Theorem 6.5 *Let P be the Riemann surface of a nonsingular irreducible polynomial $F(y, z)$ and $g: P \to \bar{\mathbb{C}}$ be a meromorphic function. Then there exists a rational function $R(y, z)$ such that $g(p) = R(y(p), z(p))$.*

Proof On the surface P consider a function $z: P \to \bar{\mathbb{C}}$ and a noncritical value $z_0 \in \bar{\mathbb{C}}$ of this function. The preimage $z^{-1}(z_0)$ consists of pairwise distinct points $\{p_1, \ldots, p_n\}$. Consider the polynomial

$$\Phi(y) = \Phi_{z_0}(y) = F(z_0, y) = (y - y(p_1)) \ldots (y - y(p_n)).$$

This is a polynomial in y whose coefficients are symmetric functions of p_i and, consequently, depend rationally on $z_0 = z(p_i)$. Besides, the polynomial $\Phi(y)$ has no multiple roots, since otherwise the pair of numbers $(y(p_i), z_0) = (y(p_j), z_0)$ would describe two different points of the surface P.

Thus, the polynomials $\Phi(y)$ and $\Phi'(y)$ are relatively prime. Therefore, there exist functions $H(y) = H_{z_0}(y)$ and $L(y) = L_{z_0}(y)$ that are polynomials in y and satisfy $H\Phi' + L\Phi = 1$. Moreover, the Euclidian algorithm, which allows one to find $H_{z_0}(y)$ and $L_{z_0}(y)$, guarantees that the coefficients of these polynomials depend rationally on z_0. Besides, $H(y(p_i))\Phi'(y(p_i)) = 1$ and $\Phi'(y(p_i)) = 1$.

Let

$$G(y) = G_{z_0}(y) = \left(\frac{g(p_1)}{y - y(p_1)} + \ldots + \frac{g(p_n)}{y - y(p_n)} \right) \Phi(y)$$

$$= \left(\frac{g(p_1)}{y - y(p_1)} + \ldots + \frac{g(p_n)}{y - y(p_n)} \right) (y - y(p_1)) \ldots (y - y(p_n)).$$

Then

$$g(p_i) = G(y(p_i)) = \frac{G(y(p_i))}{\Phi'(y(p_i))} = \frac{H(y(p_i)) G(y(p_i))}{H(y(p_i)) \Phi'(y(p_i))} = G_{z_0}(y(p_i)) H_{z_0}(y(p_i)).$$

The functions G_{z_0} and H_{z_0} are polynomials whose coefficients depend rationally on z_0. Thus, the function $R(z_0, y) = G_{z_0}(y) H_{z_0}(y)$ is rational. Besides, we have $g(p) = R(y(p), z(p))$, since the functions coincide at all preimages $z^{-1}(z_0)$ for all z_0.

Consider the algebra $\mathscr{M}$ over $\mathbb{C}$ freely generated by (multiplicative) generators y and z. Its elements can be regarded as polynomials in (y, z) with complex coefficients. Each polynomial $F(y, z)$ generates an ideal $I(F) \subset \mathscr{M}$. The previous theorem implies the following corollaries.

Corollary 6.3 *The field $\mathscr{M}(P)$ of meromorphic functions on a connected Riemann surface P is naturally isomorphic to the field $\mathscr{M}(F) = \mathscr{M}/I(F)$ where $F(y, z)$ is an irreducible polynomial whose Riemann surface is isomorphic to P. This isomorphism identifies the ideal $I(F)$ with the ideal $I(P)$ of polynomials generating the zero function on P.*

Corollary 6.4 *A nonsingular irreducible complex polynomial $F(x, y)$ generates a connected compact Riemann surface $\bar{P}_F$.*

Proof Assume that the surface $P = \bar{P}_F$ breaks into connected components P_1, P_2. Then $P_i = \bar{P}_{F_i}$ where F_1, F_2 are irreducible nonsingular polynomials, with $F_1 \neq F_2$ since F is nonsingular. As we have already proved, the field $\mathscr{M}(P_i)$ can be identified with the field $\mathscr{M}(F_i)$. The ideal $I(F_i)$ coincides with the ideal of polynomials generating the zero function on P. Therefore, $I(F) \subset I(F_i)$ and $F = G_i F_i$ where $G_i \in \mathscr{M}(F)$. Thus, $F = G F_1 F_2$ where $G \in \mathscr{M}(F)$.

Exercise 6.14 Consider connected Riemann surfaces $P_i = \bar{P}_{F_i}$ where $F_1(y, z), F_2(\tilde{y}, \tilde{z})$ are irreducible nonsingular polynomials. Then a birational change of variables $y = y(\tilde{y}, \tilde{z}), z = z(\tilde{y}, \tilde{z})$ sending $F_1(y, z)$ to $F_2(\tilde{y}, \tilde{z})$ generates a morphism $P_1 \to P_2$, and every morphism $P_1 \to P_2$ is generated by such a birational change of variables.

We have proved that the category of connected Riemann surfaces is isomorphic to the category of nonsingular irreducible polynomials $F(y, z)$ with complex coefficients whose morphisms are rational changes of variables $y = y(\tilde{y}, \tilde{z})$, $z = z(\tilde{y}, \tilde{z})$. The latter category can be studied by purely algebraic methods. Thus, many geometric results on Riemann surfaces can be restated in purely algebraic terms. In this restatement, many results remain valid when the field of complex numbers is replaced by an arbitrary algebraically closed field, for instance, the field of algebraic numbers. Results about polynomials with coefficients from the field of algebraic numbers describe properties of integers. This approach is the starting point of *algebraic number theory*. It allows one to develop number theory using geometric ideas of the theory of Riemann surfaces.

6.5 Periods of Holomorphic Differentials

According to our definitions, a meromorphic differential ω in every local chart $z \colon U \to \mathbb{C} = \{(x + iy)\}$ can be written in the form

$$\omega = f(z)\,dz = S(x, y)\,dx + T(x, y)\,dy$$

where $dz = dx + i\,dy$ and S, T are complex-valued functions.

Exercise 6.15 Show that $\omega = S(x, y)\,dx + T(x, y)\,dy$ is a closed differential 1-form (i.e., $d\omega = 0$). In particular, for every oriented segment $l \subset P$ there is a well-defined integral $\int_l \omega$, which does not change under endpoint-preserving homotopies of I.

Exercise 6.16 Let ω' be another meromorphic differential. Show that $\omega \wedge \omega' = 0$.

A *canonical basis of cycles* on a surface P of genus g is a system of simple closed oriented contours $\{a_i, b_i \mid i = 1, \ldots, g\}$ that meet at one point, have no other pairwise intersection points, and have the following intersection indices: $(a_i, a_j) = (b_i, b_j) = 0$, $(a_i, b_j) = \delta_{ij}$, where $\delta_{ij} = 1$ if $i = j$, and $\delta_{ij} = 0$ for $i \neq j$.

Theorem 6.6 *Let* ω, ω' *be closed differential* 1*-forms that are smooth on the surface* P. *Consider a canonical basis of cycles* $\mathfrak{B} = \{a_i, b_i \mid i = 1, \ldots, g\}$ *in* P. *Set*

$$A_i = \oint_{a_i} \omega, \quad B_i = \oint_{b_i} \omega, \quad A_i' = \oint_{a_i} \omega', \quad B_i' = \oint_{b_i} \omega'.$$

Then

$$\iint_P \omega \wedge \omega' = \sum_{i=1}^{g} (A_i B_i' - A_i' B_i).$$

Proof Cutting the surface P along the cycles from $\mathfrak{B}$, we obtain a $4g$-gon Γ with sides $a_1, b_1, a_1^{-1}, b_1^{-1}, a_1, b_1, a_1^{-1}, b_1^{-1}, \ldots, a_g, b_g, a_g^{-1}, b_g^{-1}$. Fix a point $p_0 \in P$ and consider the integrals $f(p) = \int_{[p_0, p]} \omega$ along paths in Γ that start at p_0 and end at p. Then $d(f\omega') = df \wedge \omega' = \omega \wedge \omega'$ and, by Stokes' theorem,

$$\iint_P \omega \wedge \omega' = \oint_{\partial \Gamma} f\omega' = \sum_{i=1}^{g} \left(\oint_{a_i} f\omega' + \oint_{a_i^{-1}} f\omega' \right) + \sum_{i=1}^{g} \left(\oint_{b_i} f\omega' + \oint_{b_i^{-1}} f\omega' \right).$$

On the other hand, for points $p_i \in a_i$, $\tilde{p}_i \in a_i^{-1}$ corresponding to the same point $p \in P$, we have $f(\tilde{p}_i) - f(p_i) = \oint_{b_i} \omega = B_i$. Thus,

$$\oint_{a_i} f(p_i)\omega'(p_i) + \oint_{a_i^{-1}} f(\tilde{p}_i)\omega'(\tilde{p}_i)$$

$$= \oint_{a_i} f(p_i)\omega'(p_i) - \oint_{a_i} (f(p_i) + B_i)\omega'(p_i) = -B_i \oint_{a_i} \omega'(p_i) = -B_i A_i'.$$

In a similar way, $\oint_{b_i} f(p_i)\omega'(p_i) + \oint_{b_i^{-1}} f(\tilde{p}_i)\omega'(\tilde{p}_i) = A_i B_i'$. Thus,

$$\iint_P \omega \wedge \omega' = \sum_{i=1}^{g}(A_i B_i' - A_i' B_i).$$

Corollary 6.5 *Let* ω *and* ω' *be holomorphic differentials and* $A_i = \oint_{a_i} \omega$, $B_i = \oint_{b_i} \omega$,
$A_i' = \oint_{a_i} \omega'$, $B_i' = \oint_{b_i} \omega'$. *Then* $\sum_{i=1}^{g}(A_i B_i' - A_i' B_i) = 0$. *If, besides,* $\omega \neq 0$, *then*

$$\Im\left(\sum_{i=1}^{g}(A_i \bar{B}_i)\right) < 0.$$

In particular, $\omega = 0$ *if* $A_i = 0$ *for all* i *or* A_i, $B_i \in \mathbb{R}$.

Proof The first assertion follows from the equality $\omega \wedge \omega' = 0$. The differential $\bar{\omega} = \bar{f}(\bar{z})\,d\bar{z}$ complex conjugate to $\omega = f(z)\,dz$ is also closed, with $\bar{A}_i = \oint_{a_i} \bar{\omega}$, $\bar{B}_i = \oint_{b_i} \bar{\omega}$. Hence

$$\Im\left(\sum_{i=1}^{g}(A_i \bar{B}_i)\right) = -\frac{i}{2}\sum_{i=1}^{g}(A_i \bar{B}_i - \bar{A}_i B_i) = -\frac{i}{2}\iint_P \omega \wedge \bar{\omega} = -\iint_P |f|^2\, dx \wedge dy < 0.$$

Exercise 6.17 Show that the dimension of the space of holomorphic differentials on a Riemann surface of genus g is equal to g. Moreover, for every canonical basis of cycles there exists a unique basis of holomorphic differentials $\{\omega_1, \ldots, \omega_g\}$ such that $\int_{a_i} \omega_j = 2\pi i\,\delta_{ij}$.

A *period of a meromorphic differential* ω is the integral of ω along a closed contour. Consider the basis of holomorphic differentials $\{\omega_1, \ldots, \omega_g\}$ from the previous exercise. The arising matrix $B_{ij} = \int_{b_i} \omega_j$ is called the *period matrix* on P corresponding to the canonical basis of cycles $\{a_i, b_i \mid i = 1, \ldots, g\}$.

Theorem 6.7 *The period matrix corresponding to a canonical basis of cycles* $\{a_i, b_i \mid i = 1, \ldots, g\}$ *is symmetric, and its real part is negative definite.*

Proof Set $\omega = \omega_k$, $\omega' = \omega_j$ and define A_i, B_i, A_i', B_i' as in Theorem 6.6. Then, by Corollary 6.5, we have

$$0 = \sum_{i=1}^{g}(A_i B_i' - A_i' B_i) = \sum_{i=1}^{g}(2\pi i\,\delta_{ik}B_{ij} - 2\pi i\,\delta_{ij}B_{ik}) = 2\pi i(B_{kj} - B_{jk}).$$

Thus, $B_{kj} = B_{jk}$.

Consider a linear combination $\omega = \sum_{k=1}^{g} x_k \omega_k$ with real coefficients and set $A_i = \oint_{a_i} \omega$, $B_i = \oint_{b_i} \omega$. Then $A_i = 2\pi i \, x_i$, $B_i = \sum_{k=1}^{g} x_k B_{ik}$ and, by Corollary 6.5, we have

$$0 > \Im\left(\sum_{i=1}^{g}(A_i \bar{B}_i)\right) = \Im\left(\sum_{k=1}^{g} 2\pi i \, x_i \sum_{k=1}^{g} x_k B_{ik}\right) = 2\pi \sum_{i,k}(\Re B_{ij})x_i x_k.$$

6.6 Riemann's Bilinear Relations

According to our definitions, a meromorphic differential ω in every local chart $z: U \to \mathbb{C}$ where $z(p) = 0$ can be written in the form

$$\omega = f(z)\,dz = (c_k z^k + c_{k+1} z^{k+1} + \ldots)\,dz$$

where k is an arbitrary integer. Consider a small contour c around the point p for which the contour $z(c)$ is oriented counterclockwise. The number

$$\mathrm{Res}_p \, \omega = \frac{1}{2\pi i} \oint_c \omega = c_{-1}$$

is called the *residue of the differential ω at the point p*.

Exercise 6.18 Show that a meromorphic differential ω on a compact surface has finitely many poles, and the sum of the residues over all poles vanishes. In particular, if ω has a unique pole, then the order of this pole is greater than 1.

Theorem 6.8 *Let ω be a holomorphic differential on a Riemann surface P and ω' be a meromorphic differential on P. Let $\mathfrak{B} = \{a_i, b_i \mid i = 1, \ldots, g\}$ be a canonical basis of cycles on P. Set*

$$A_i = \oint_{a_i} \omega, \quad B_i = \oint_{b_i} \omega, \quad A_i' = \oint_{a_i} \omega', \quad B_i' = \oint_{b_i} \omega'.$$

Assume that ω' has a unique pole and

$$\omega' = \left(\frac{1}{z^n} + \sum_{i=0}^{\infty} a_i z^i\right) dz$$

in some local chart $z: U \to \mathbb{C}$. *Consider the representation*

$$\omega = \left(\sum_{j=0}^{\infty} c_j z^j \right) dz$$

of ω *in this local chart. Then*

$$\sum_{i=1}^{g} (A_i B_i' - A_i' B_i) = 2\pi i \frac{c_{n-2}}{n-1}.$$

Proof Consider a small contour γ around the pole of ω'. Cutting the surface $P \setminus \gamma$ along the cycles from $\mathfrak{B}$, we obtain a $4g$-gon Γ with sides

$$a_1, b_1, a_1^{-1}, b_1^{-1}, a_1, b_1, a_1^{-1}, b_1^{-1}, \ldots, a_g, b_g, a_g^{-1}, b_g^{-1}$$

with the hole γ. Fix a point $p_0 \in \Gamma$ and consider the integrals $f(p) = \int_{[p_0, p]} \omega$ along paths in Γ that start at p_0 and end at p. Then $d(f\omega') = df \wedge \omega' = \omega \wedge \omega'$ and, by Stokes' theorem,

$$0 = \iint_{P} \omega \wedge \omega' = \oint_{\partial \Gamma} f\omega' = \sum_{i=1}^{g} \left(\oint_{a_i} f\omega' + \oint_{a_i^{-1}} f\omega' \right) + \sum_{i=1}^{g} \left(\oint_{b_i} f\omega' + \oint_{b_i^{-1}} f\omega' \right) - \oint_{\gamma} f\omega'.$$

The first two terms can be found as in the proof of Theorem 6.6. They are equal to $-\sum_{i=1}^{g} B_i A_i'$ and $\sum_{i=1}^{g} A_i B_i'$, respectively. The last integral is a residue of the holomorphic differential $f\omega'$, therefore, it is equal to $2\pi i \frac{c_{n-2}}{n-1}$.

Theorem 6.9 *Let* ω *be a holomorphic differential on a Riemann surface* P *of genus g and* ω' *be a meromorphic differential on* P. *Let* $\mathfrak{B} = \{a_i, b_i \mid i = 1, \ldots, g\}$ *be a canonical basis of cycles on* P. *Set*

$$A_i = \oint_{a_i} \omega, \quad B_i = \oint_{b_i} \omega, \quad A_i' = \oint_{a_i} \omega', \quad B_i' = \oint_{b_i} \omega'.$$

Choose a point $p_0 \in P \setminus \mathfrak{B}$ *and set* $f(p) = \int_{l_p} \omega$ *where* l_p *is a path in* $P \setminus \mathfrak{B}$ *that starts at* p_0 *and ends at* p. *If* ω' *has only simple poles* p_k, *then*

$$\sum_{i=1}^{g} (A_i B_i' - A_i' B_i) = 2\pi i \sum_{k} f(p_k) \operatorname{Res}_{p_k}(\omega').$$

Proof Repeating the beginning of the proof of the previous theorem with obvious modifications, we obtain

$$0 = \sum_{i=1}^{g} (A_i B_i' - A_i' B_i) - \sum_k \oint_{\gamma_k} f\omega'$$

where γ_k are small contours around p_k. On the other hand,

$$\oint_{\gamma_k} f\omega' = 2\pi i f(p_k) \operatorname{Res}_{p_k}(\omega'),$$

since the poles at p_i are simple.

Corollary 6.5 and Theorems 6.8, 6.9 are called *Riemann's bilinear relations*.

Chapter 7
The Riemann–Roch Theorem and Theta Functions

7.1 Divisors

A finite formal linear combination

$$D = \sum_{i=1}^{k} n_i p_i$$

of points $p_i \in P$ of a Riemann surface P with integer coefficients $n_i \in \mathbb{Z}$ is called a *divisor* on P. The set of divisors is a module over the ring of integers $\mathbb{Z}$. The zero element of this module is $D = \varnothing$, a sum with no terms.

The *degree*

$$\deg D = \sum_{i=1}^{k} n_i$$

of a divisor D is a linear functional on this module. A divisor is said to be *positive* (notation: $D > 0$) if all its coefficients n_i are positive. We will also write $D_1 > D_2$ if $D_1 - D_2 > 0$.

If h is a meromorphic function or a differential with the set of zeros $p_1, \ldots, p_s$ and the set of poles $q_1, \ldots, q_t$, then the *divisor* of h is

$$(h) = \sum_{i=1}^{s} n_i p_i - \sum_{i=1}^{t} m_i q_i$$

where n_i (respectively, m_i) is the order of zero (respectively, pole) of h at the point p_i (respectively, q_i).

© Springer Nature Switzerland AG 2019
S. M. Natanzon, *Complex Analysis, Riemann Surfaces and Integrable Systems*,
Moscow Lectures 3, https://doi.org/10.1007/978-3-030-34640-9_7

The divisor of a meromorphic function is called a *principal divisor*. Divisors that differ by a principal divisor are said to be *linearly equivalent*.

Exercise 7.1 Show that the degrees of linearly equivalent divisors coincide.

A divisor linearly equivalent to a positive divisor is said to be *effective*.

The divisors of meromorphic differentials form a linear equivalence class $\mathcal{K}$ called the *canonical class*.

Exercise 7.2 Using the Riemann–Hurwitz theorem, show that $\deg(\mathcal{K}) = 2g - 2$.

With a divisor D we associate the vector spaces

$$R(D) = \{f \text{ is a meromorphic function such that } (f) \geq D\}$$

and

$$I(D) = \{\omega \text{ is a meromorphic differential such that } (\omega) \geq D\}.$$

Let $r(D) = \dim R(D)$ and $i(D) = \dim I(D)$.

Example 7.1 If $D > 0$, then $R(D) = \varnothing$ and $r(D) = 0$. If $D = \varnothing$, then the set $R(D)$ consists of the constant functions, $r(D) = 1$, and $i(D) = g$.

Theorem 7.1 *For any divisor D and any meromorphic function f, we have*

$$i(D) = r(D - \mathcal{K}) \quad and \quad i(D + (f)) = i(D), \quad r(D + (f)) = r(D).$$

Proof Let ω be a meromorphic differential. Then the correspondence $\omega' \mapsto \frac{\omega'}{\omega}$ establishes an isomorphism between $I(D)$ and $R(D - (\omega))$, which implies that $i(D) = r(D - \mathcal{K})$. The correspondence $f' \mapsto f'f$ establishes an isomorphism between $R(D)$ and $R(D + (f))$. In a similar way, we obtain that $i(D+(f)) = i(D)$.

7.2 The Riemann–Roch Theorem

The *Riemann–Roch theorem states that*

$$r(-D) = \deg D + i(D) - g + 1$$

for all divisors D.

Let us rewrite it in a symmetric form.

Lemma 7.1 *The equality $r(-D) = \deg D + i(D) - g + 1$ holds if and only if*

$$r(-D) + \frac{1}{2}\deg(-D) = r(D - \mathcal{K}) + \frac{1}{2}\deg(D - \mathcal{K}).$$

Proof Let $r(-D) = \deg D + i(D) - g + 1$. Then, by Theorem 7.1 and Exercise 7.2, we obtain

$$r(-D) + \frac{1}{2}\deg(-D) = \deg D + i(D) - g + 1 + \frac{1}{2}\deg(-D)$$

$$= \frac{1}{2}\deg(D) + i(D) - \frac{1}{2}\deg(\mathcal{K}) = r(D - \mathcal{K}) + \frac{1}{2}\deg(D - \mathcal{K}).$$

The converse can be proved in a similar way.

We know already (Theorem 6.2) that the Riemann–Roch theorem holds for $D = \varnothing$, since $r(\varnothing) = 1$ and $i(\varnothing) = g$. Let us prove another special case of the Riemann–Roch theorem.

Theorem 7.2 *If $D > 0$, then $r(-D) = \deg D + i(D) - g + 1$.*

Proof Fix a canonical basis of cycles $\{a_i, b_i \mid i = 1, \ldots, g\}$ on the Riemann surface. Let $D = \sum_{k=1}^{m} n_k p_k$ with $n_k > 0$. By Theorem 6.2, for every point p_k and every $j > 1$ there exist a local chart (U, z) and a meromorphic differential φ_k^j that in this chart has the form $\varphi_k^j = \left(\frac{1}{z^j} + \sum_{s=0}^{\infty} a_s z^s\right) dz$ and has no other poles on the whole surface. In view of Exercise 6.17, we may assume that $\oint_{a_i} \varphi_k^j = 0$ for all i. Set $B_{kl}^j = \oint_{b_l} \varphi_k^j$ (here the index k corresponds to the point, l to the integration contour, and j to the degree of the pole of the differential). These numbers constitute the matrix

$$B = \begin{pmatrix} B_{11}^2 & B_{11}^3 & \cdots & B_{11}^{n_1+1} & B_{21}^2 & \cdots & B_{m1}^{n_m+1} \\ B_{12}^2 & B_{12}^3 & \cdots & B_{12}^{n_1+1} & B_{22}^2 & \cdots & B_{m2}^{n_m+1} \\ & & & \cdots\cdots\cdots & & & \\ B_{1g}^2 & B_{1g}^3 & \cdots & B_{1g}^{n_1+1} & B_{2g}^2 & \cdots & B_{mg}^{n_m+1} \end{pmatrix},$$

in which rows represent integration contours; and columns, degrees of poles at every point.

We will be interested in the solutions $\{c_{-j}^k\}$ to the system of equations

$$\sum_{k=1}^{m} \sum_{j=2}^{n_k+1} c_{-j}^k B_{kl}^j = 0 \quad (l = 1, \ldots, g).$$

As we learn from linear algebra, the dimension of the space of such solutions is equal to $\deg D - \rho$, where $\deg D = \sum_{k=1}^{m} n_k$ is the number of unknowns and ρ is the rank of the matrix B.

Let us prove that every function from $R(-D)$ yields a solution to the system. In the chosen local charts in neighborhoods of the points p_k, the differential of a function $f \in R(-D)$ can be written in the form

$$df = \left(\sum_{j=-n_k-1}^{\infty} c_j^k z^j \right) dz = \left(\sum_{j=2}^{n_k+1} c_{-j}^k z^{-j} + \sum_{j=0}^{\infty} c_j^k z^j \right) dz,$$

since $c_{-1}^k = 0$. The differential $\varphi = df - \sum_{k=1}^{m} \sum_{j=2}^{n_k+1} c_{-j}^k \varphi_k^j$ is holomorphic on the whole surface, and $\oint_{a_i} \varphi = 0$ for all i. Thus, by Corollary 6.5, we have $\varphi = 0$, i.e.,

$$df = \sum_{k=1}^{m} \sum_{j=2}^{n_k+1} c_{-j}^k \varphi_k^j.$$

Therefore,

$$\sum_{k=1}^{m} \sum_{j=2}^{n_k+1} c_{-j}^k B_{kl}^j = \oint_{b_l} \left(\sum_{k=1}^{m} \sum_{j=2}^{n_k+1} c_{-j}^k \varphi_k^j \right) = \oint_{b_l} df = 0 \quad \text{for } l = 1, \ldots, g.$$

Functions that differ by a constant yield the same solution.

The converse is also true. To every nonzero collection $\{c_{-j}^k\}$ there corresponds a meromorphic differential

$$\omega = \sum_{k=1}^{m} \sum_{j=2}^{n_k+1} c_{-j}^k \varphi_k^j.$$

All periods of ω are zero if $\{c_{-j}^k\}$ is a solution to the system. Therefore, $\omega = df$ where the function $f = \int_{p_0}^{p} \omega$ belongs to $R(-D)$ and is defined up to an additive constant.

Thus, the space $R(-D)$ is spanned by the constants and the solutions to our system. Hence, $\dim R(-D) = \deg D - \rho + 1$ and $-\rho = r(-D) - \deg D - 1$.

Now let us represent the matrix B in terms of holomorphic differentials. Consider a basis $\{\omega_1, \ldots, \omega_g\}$ of the space of holomorphic differentials for which $\oint_{a_i} \omega_l = \delta_{il}$. Consider the representations

$$\omega_l = \left(\sum_{j=0}^{\infty} \alpha_{lj}^k z^j \right) dz$$

of these differentials in the local charts in neighborhoods of the points p_k chosen earlier. Riemann's bilinear relation (Theorem 6.8) applied to the pair of differentials ω_l and φ_k^j implies that

$$2\pi i \frac{\alpha_{l(j-2)}^k}{j-1} = \sum_{i=1}^{g}(\delta_{il}B_{ki}^j - 0) = B_{kl}^j \,.$$

Thus,

$$B = 2\pi i \begin{pmatrix} \alpha_{10}^1 & \dfrac{\alpha_{11}^1}{2} & \cdots & \dfrac{\alpha_{1n_1-1}^1}{n_1} & \alpha_{10}^2 & \cdots & \dfrac{\alpha_{1n_m-1}^m}{n_m} \\[2mm] \alpha_{20}^1 & \dfrac{\alpha_{21}^1}{2} & \cdots & \dfrac{\alpha_{2n_1-1}^1}{n_1} & \alpha_{20}^2 & \cdots & \dfrac{\alpha_{2n_m-1}^m}{n_m} \\[2mm] & & \cdots\cdots\cdots & & & & \\[2mm] \alpha_{g0}^1 & \dfrac{\alpha_{g1}^1}{2} & \cdots & \dfrac{\alpha_{gn_1-1}^1}{n_1} & \alpha_{g0}^2 & \cdots & \dfrac{\alpha_{gn_m-1}^m}{n_m} \end{pmatrix}.$$

A row of this matrix consists of the expansion coefficients of one of the differentials ω_l at all points of the divisor. A column represents the expansion coefficients of all differentials ω_l of the same degree at the same point. Therefore, a differential $\sum_{l=1}^{m} d_l\omega_l$ belongs to $I(D)$ if and only if the linear combination of the rows of B with the coefficients $\{d_l\}$ vanishes. The dimension of the vector space of such collections $\{d_l\}$ is equal to $g - \rho$. Thus, $i(D) = g - \rho = g + r(-D) - \deg D - 1$.

This theorem, together with Theorem 7.1, implies the following result.

Corollary 7.1 *If D is an effective divisor, then $r(-D) = \deg D + i(D) - g + 1$.*

Theorem 7.3 *For every divisor D, we have $r(-D) = \deg D + i(D) - g + 1$.*

Proof If $r(-D) > 0$, then $(f) + D > 0$ for $f \in R(-D)$. Hence D is an effective divisor and the Riemann–Roch theorem holds by Corollary 7.1. By Lemma 7.1, the theorem also holds in the case $r(D - \mathcal{K}) > 0$.

Now let $r(-D) = 0$ and $r(D - \mathcal{K}) = 0$. Represent the divisor in the form $D = D_+ - D_-$ where D_+, D_- are positive or empty. Assume that $\deg D \geq g$. Then, as we have already proved,

$$r(-D_+) \geq \deg D_+ - g + 1 = \deg D + \deg D_- - g + 1 \geq 1 + \deg D_-.$$

Thus, the space $R(-D_+)$ contains $1 + \deg D_-$ linearly independent functions. Considering their linear combinations, we can find a function $f \in R(-D_+)$ whose divisor of zeros is D_-, i.e., a function $f \in R(-D)$. This contradicts the assumption $r(-D) = 0$; therefore, $\deg D < g$. By Lemma 7.1, it follows that $\deg(\mathcal{K} - D) < g$ and, in particular, $\deg(D) > 2g - 2 - g = g - 2$.

Thus, $\deg(D) = g - 1$; since $r(-D) = 0$ and $i(D) = r(D - \mathcal{K}) = 0$, this implies the Riemann–Roch theorem.

7.3   Weierstrass Points

The Riemann–Roch theorem allows one to estimate the number of linearly independent functions with prescribed singularities. By the Riemann–Roch theorem, $r(-D) > 1$ for $\deg D > g$. Let us find an upper bound on the number of linearly independent functions.

Lemma 7.2 *If $g > 0$, then $i(p) = g - 1$ for every point $p \in P$. If $D > 0$ and $g = 0$, then $r(-D) = \deg D + 1$. If $D > 0$ and $g > 0$, then $r(-D) < \deg D + 1$.*

Proof If $i(p) \geq g$, then $r(-p) \geq 1 + g - g + 1 = 2$ and there exists a function with a unique simple pole, which is possible only if $g = 0$ (Exercise 6.18). Thus, $i(p) < g$ for $g > 0$. On the other hand, $i(p) = r(-p) - 1 + g - 1 \geq g - 1$ and, therefore, $i(p) = g - 1$ for $g > 0$.

If $g = 0$ and $D > 0$, then $i(D) = 0$, whence $r(-D) = \deg D + 1$. Now let $g > 0$, $D > 0$, and $p \in D$. Then $I(D) \subset I(p)$ and, therefore, $i(D) \leq i(p) = g - 1$, whence $r(-D) < \deg D + 1$.

Theorem 7.4 *On a surface of genus $g > 0$, for every $j = 1, \ldots, g$ there exist pairwise distinct points $p_1, \ldots, p_j$ such that $i\left(\sum_{k=1}^{j} p_k\right) = g - j$. Besides, on a surface of genus $g > 0$ there exist g pairwise distinct points $p_1, \ldots, p_g$ such that $r\left(-\sum_{k=1}^{g} p_k\right) = 1$.*

Proof We prove the first assertion by induction on j. For $j = 1$, it is proved in Lemma 7.2. Assume that the assertion holds for $j = n < g$, i.e., there exist pairwise distinct points $p_1, \ldots, p_n$ such that $i\left(\sum_{k=1}^{n} p_k\right) = g - n$. Then there exist a nonzero differential $\omega \in I\left(\sum_{k=1}^{n} p_k\right)$ and a point p_{n+1} such that $\omega(p_{n+1}) \neq 0$. Therefore, ω does not belong to $I\left(\sum_{k=1}^{n+1} p_k\right)$. Hence, $i\left(\sum_{k=1}^{n+1} p_k\right) < g - n$. On the other hand,

$$i\left(\sum_{k=1}^{n+1} p_k\right) = r\left(-\sum_{k=1}^{n+1} p_k\right) + g - 1 - (n+1) \geq g - (n+1).$$

Thus, $i\left(\sum_{k=1}^{n+1} p_k\right) = g - (n+1)$. The second assertion of the theorem follows from the first one and the Riemann–Roch theorem.

One can see from the proof of the last theorem that the property $r\left(-\sum_{k=1}^{g} p_k\right) = 1$ holds for every collection of points $p_1, \ldots, p_g$ in general position. Arbitrary collections of points for which $r\left(-\sum_{k=1}^{g} p_k\right) > 1$ will be discussed below. Here we restrict ourselves to divisors of this type concentrated at a single point.

A point p on a surface of genus g is called a *Weierstrass point* if $r(-gp) > 1$.

Lemma 7.3 *Let z be a local chart on a Riemann surface P such that $z(p) = 0$, and let $\omega_i = \varphi_i(z)\, dz \ (i = 1, \ldots, g)$ be a basis of the space of holomorphic differentials on P. Set*

$$W(z) = \det \begin{pmatrix} \varphi_1(z) & \varphi_1'(z) & \cdots & \varphi_1^{(g-1)}(z) \\ & \cdots \cdots \cdots & \\ \varphi_g(z) & \varphi_g'(z) & \cdots & \varphi_g^{(g-1)}(z) \end{pmatrix}.$$

Then p is a Weierstrass point if and only if $W(0) = 0$.

Proof By the Riemann–Roch theorem, $r(-gp) > 1$ if and only if $i(gp) > 0$, i.e., there exists a holomorphic differential with a zero of order g at p. Such a differential $\omega = \sum_{i=1}^{g} \lambda_i \omega_i$ exists if and only if $W(0) = 0$.

Exercise 7.3 Show that under a change of chart $z = z(u)$, the function W changes as follows:

$$\tilde{W}(u) = \left(\frac{dz}{du}\right)^N W(z) \quad \text{where} \quad N = \frac{g(g+1)}{2}.$$

The multiplicity of the zero of the function W at 0 is called the *weight* of the Weierstrass point.

Exercise 7.4 Show that the weight of a Weierstrass point does not depend on the choice of a basis $\{\varphi_i\}$.

Theorem 7.5 *The total weight of Weierstrass points on a Riemann surface of genus g is equal to $(g - 1)g(g + 1)$.*

Proof By Exercise 7.4, the function $f(z) = \frac{W(z)}{(\varphi_1(z))^N}$ does not depend on the choice of a local chart and, consequently, generates a meromorphic function on P. Its degree is equal to the degree of the divisor of zeros and the degree of the divisor of poles of f. The degree of the divisor of zeros is the total weight of Weierstrass points. The degree of the divisor of poles is equal to the degree of the divisor of zeros of the tensor $(\varphi_1(z))^N$, which, by Exercises 7.2 and 7.3, is equal to $(2g - 2)N = (g - 1)g(g + 1)$.

Exercise 7.5 (Hurwitz' Theorem) Using Theorem 7.5 and the Riemann–Hurwitz theorem, show that the order of the group of automorphisms of a Riemann surface of genus $g > 1$ does not exceed $84(g - 1)$. (*Hint.* Consider the quotient of the surface by the action of the group of automorphisms (which is finite by Theorem 7.5) and apply the Riemann–Hurwitz formula replacing the sum over critical points by a sum over critical values.)

An integer a is called a *gap* at a point p on a Riemann surface P if there is *no* meromorphic function on P whose divisor of poles is ap.

Theorem 7.6 *On a surface of genus $g > 0$, at every point there are exactly g gaps; all of them do not exceed $2g - 1$.*

Proof The number $r(-kp)$ does not decrease as k grows, and, by the Riemann–Roch theorem, we have $r(-(k+1)p) - r(-kp) = i((k+1)p) - i(kp) + 1 \leq 1$. Besides, $r(-p) = 1$ and $r(-(2g-1)p) = 2g-1+i((2g-1)p)-g+1 = g$, since $\deg(\mathcal{K}) = 2g - 2$. Thus, in the interval $1 \leq k \leq 2g-1$ the function $r(-kp)$ jumps exactly $g - 1$ times and, consequently, does not jump exactly g times. If $k \geq 2g$, then, by the Riemann–Roch theorem, $r(-kp) = k - g + 1 > 1$, i.e., functions with the divisor of poles kp do exist.

Exercise 7.6 Show that the weight of a Weierstrass point equals $\sum_{i=1}^{g} (a_i - i)$ where $a_1 < a_2 < \ldots < a_g$ are its gaps.

Exercise 7.7 Find the Weierstrass points and their gaps for a hyperelliptic Riemann surface.

Exercise 7.8 Show that the Riemann surface of genus $g > 1$ has at least $2g + 2$ Weierstrass points.

7.4 Abelian Tori and Theta Functions

A symmetric $g \times g$ matrix $B = (B_{ij})$ with negative definite real part is called a *Riemann matrix*.

Lemma 7.4 *Let $B = (B_{ij})$ be a Riemann matrix. Then the vectors*

$$2\pi i \, e_k = 2\pi i \begin{pmatrix} 0 \\ \cdots \\ 1 \\ \cdots \\ 0 \end{pmatrix}, \quad f_l = Be_l = \begin{pmatrix} B_{l1} \\ \cdots \\ B_{lg} \end{pmatrix}, \quad k, l = 1, \ldots, g,$$

are linearly independent over $\mathbb{R}$.

Proof Taking the real part of both sides of the equation $2\pi i \sum_{k=1}^{g} \lambda_k e_k + \sum_{k=1}^{g} \mu_k f_k = 0$

(with $\lambda_k, \mu_k \in \mathbb{R}$), we obtain $\Re B\left(\sum_{k=1}^{g} \mu_k e_k\right) = 0$, which, since $\Re B$ is nondegen-

erate, implies that $\sum_{k=1}^{g} \mu_k e_k = 0$. Therefore, $\mu_1 = \ldots = \mu_g = 0$, and hence $\lambda_1 = \ldots = \lambda_g = 0$.

One can easily show that there are no holomorphic functions on $\mathbb{C}^g$ with $2g$ linearly independent periods. In complex analysis, the role of such functions is played by theta functions, which most closely resemble them.

The *theta function associated with a Riemann matrix B* is the function $\theta : \mathbb{C}^g \to \mathbb{C}$ given by

$$\theta(z) = \theta(z|B) = \sum_{N \in \mathbb{Z}^g} \exp\left(\frac{1}{2}\langle BN, N\rangle + \langle N, z\rangle\right)$$

where $z = (z_1, \ldots, z_g)$, $N = (N_1, \ldots, N_g)$, $\langle N, z\rangle = \sum_{i=1}^{g} N_i z_i$, $\langle BN, N\rangle = \sum_{i,j=1}^{g} B_{ij} N_i N_j$.

Lemma 7.5 *The series $\theta(z)$ absolutely converges on every compact subset of $\mathbb{C}^n$.*

Proof Consider the greatest eigenvalue $-b < 0$ of the matrix $\Re B$. Then we have $\Re(\langle BN, N\rangle) \le -b \langle N, N\rangle$ and, therefore,

$$\left|\exp\left\{\frac{1}{2}\langle BN, N\rangle + \langle N, z\rangle\right\}\right| \le \exp\left\{-\frac{b}{2}\sum_{i=1}^{g} N_i^2 + \sum_{i=1}^{g} N_i \tilde{z}_i\right\}$$

$$= \prod_{i=1}^{g} \exp\left\{-\frac{b}{2}N_i^2 + N_i \tilde{z}_i\right\} \quad \text{where } \tilde{z} = \Re z.$$

Thus,

$$|\theta(z)| = \left|\sum_{N \in \mathbb{Z}^g} \exp\left\{\frac{1}{2}\langle BN, N\rangle + \langle N, z\rangle\right\}\right|$$

$$\le \sum_{N \in \mathbb{Z}^g}\left(\prod_{i=1}^{g} \exp\left\{-\frac{b}{2}N_i^2 + N_i \tilde{z}_i\right\}\right) = \prod_{i=1}^{g}\left(\sum_{n=-\infty}^{\infty} \exp\left\{-\frac{b}{2}n^2 + n\tilde{z}_i\right\}\right).$$

Therefore,

$$|\theta(z)| \le \Big(\sum_{n=-\infty}^{\infty} \exp\Big\{-\frac{b}{2}n^2 + cn\Big\}\Big)^g = \text{const}\Big(\sum_{n=-\infty}^{\infty} \exp\Big\{-\frac{b}{2}(n-\frac{c}{b})^2\Big\}\Big)^g$$

for $|\tilde{z}_i| \le c$.

The convergence of the series

$$\sum_{n=-\infty}^{\infty} \exp\Big\{-\frac{b}{2}(n-\frac{c}{b})^2\Big\}$$

follows from the convergence of the integral

$$\int_{-\infty}^{\infty} \exp\Big\{-\frac{b}{2}(x-\frac{c}{b})^2\Big\}\,dx,$$

which is equivalent to the convergence of the integral

$$\int_{-\infty}^{\infty} \exp\{-x^2\}\,dx.$$

Thus, the series $\theta_B(z)$ uniformly converges on the set

$$\{z \in \mathbb{C}^n \mid |\Re z_i| \le c\} \quad \text{for every } c.$$

Lemma 7.6 *We have*

$$\theta(z + 2\pi i e_k) = \theta(z); \quad \theta(z + f_k) = \exp\Big(-\frac{1}{2}B_{kk} - z_k\Big)\theta(z).$$

Proof The first equation is obvious. Let us prove the second one. Set $N = M - e_k$. Then

$$\theta(z + f_k) = \sum_{N \in \mathbb{Z}^g} \exp\Big(\frac{1}{2}\langle BN, N\rangle + \langle N, z + f_k\rangle\Big)$$

$$= \sum_{N \in \mathbb{Z}^g} \exp\Big(\frac{1}{2}\langle B(M-e_k), (M-e_k)\rangle + \langle(M-e_k), z + f_k\rangle\Big)$$

$$= \sum_{M \in \mathbb{Z}^g} \exp\Big(\frac{1}{2}\langle BM, M\rangle - \langle M, Be_k\rangle + \frac{1}{2}\langle Be_k, e_k\rangle$$

$$+ \langle M, z\rangle + \langle M, f_k\rangle - \langle e_k, z\rangle - \langle e_k, f_k\rangle\Big)$$

$$= \exp\left(-\frac{1}{2} B_{kk} - z_k\right) \sum_{M \in \mathbb{Z}^g} \exp\left(\frac{1}{2} \langle BM, M \rangle + \langle M, z \rangle\right)$$

$$= \exp\left(-\frac{1}{2} B_{kk} - z_k\right) \theta(z).$$

Exercise 7.9 Show that

$$\theta(z + 2\pi i N + BM|B) = \exp\left(-\frac{1}{2} \langle BM, M \rangle - \langle M, z \rangle\right) \theta(z|B)$$

for any integer vectors $N, M \in \mathbb{Z}^g$.

The lattice

$$\Gamma = \{2\pi i N + BM \mid N, M \in \mathbb{Z}^g\} \subset \mathbb{C}^g = \mathbb{R}^{2g}$$

is called the *lattice generated by the Riemann matrix B*. The rank of Γ is equal to $2g$ by Lemma 7.4. The quotient space $T^{2g} = \mathbb{C}^g/\Gamma$ is called an *Abelian torus*. Denote by $J_B : \mathbb{C}^g \to T^{2g}$ the natural projection. For vectors $V_1, V_2 \in \mathbb{C}^g$, we will also write $V_1 \equiv V_2$ if $J_B(V_1) = J_B(V_2)$.

One can show that an Abelian torus is an algebraic variety, i.e., can be defined by algebraic equations in some projective space. Moreover, one can prove that every algebraic torus corresponds to a Riemann matrix. It is exactly theta functions that determine an embedding of the torus T^{2g} as an algebraic variety [6, Sec. 2.6].

Different Riemann matrices may define the same Abelian torus.

Exercise 7.10 Show that Riemann matrices B and B' generate the same Abelian torus if and only if there exists a matrix of the form $\begin{pmatrix} a & b \\ c & d \end{pmatrix}$ such that

$$B' = 2\pi i (aB + 2\pi i b)(cB + 2\pi i d)^{-1}.$$

There are generalizations of theta functions which are important in applications, namely, *theta functions with characteristics* $\alpha, \beta \in \mathbb{R}^g$:

$$\theta[\alpha, \beta](z|B) = \sum_{N \in \mathbb{Z}^g} \exp\left(\frac{1}{2} \langle B(N+\alpha), N+\alpha \rangle + \langle z + 2\pi i \beta, N+\alpha \rangle\right), \quad \alpha, \beta \in \mathbb{R}^g.$$

Exercise 7.11 Show that

$$\theta[\alpha, \beta](z + 2\pi i N + BM|B)$$

$$= \exp\left(-\frac{1}{2} \langle BM, M \rangle - \langle z, M \rangle + 2\pi i \left(\langle \alpha, N \rangle - \langle \beta, M \rangle\right)\right) \theta[\alpha, \beta](z|B).$$

Theta functions with characteristics α, β can be expressed in terms of ordinary theta functions.

Exercise 7.12 Show that

$$\theta[\alpha, \beta](z|B) = \exp\left(\frac{1}{2}\langle B\alpha, \alpha\rangle + \langle\alpha, z + 2\pi i\beta\rangle\right)\theta(z + B\alpha + 2\pi i\beta).$$

Exercise 7.13 Show that the transformation of Riemann matrices from Exercise 7.10 generates the transformation

$$\theta[\alpha', \beta'](z'|B') = \text{const} \cdot \sqrt{M} \exp\left\{\frac{1}{2}\sum_{i\leq j} z_i z_j \frac{\partial \ln \det M}{\partial B_{ij}}\right\}\theta[\alpha, \beta](z|B),$$

where $M = cB + 2\pi i d$, $z = \frac{1}{2\pi i}z'M$, and $[\alpha', \beta'] = [\alpha, \beta]\begin{pmatrix} d & -b \\ -c & a \end{pmatrix} + \begin{pmatrix} cd^t & 0 \\ 0 & ab^t \end{pmatrix}$.

Of most importance are theta functions with characteristics whose coordinates are equal to 0 and $\frac{1}{2}$. Such characteristics are called *semiperiods*. They are said to be *even* or *odd* depending on the parity of the number $4\langle\alpha, \beta\rangle$.

Exercise 7.14 Show that the parity of a semiperiod $[\alpha, \beta]$ coincides with the parity of the corresponding theta function $\theta[\alpha, \beta]$. Find the number of even and odd semiperiods.

By a theta function of order n with characteristics $[\alpha, \beta]$ one means a holomorphic function on $\mathbb{C}^g$ satisfying the condition

$$\theta_n[\alpha, \beta](z + 2\pi i N + BM|B)$$
$$= \exp\left(-\frac{n}{2}\langle BM, M\rangle - n\langle z, M\rangle + 2\pi i\left(\langle\alpha, N\rangle - \langle\beta, M\rangle\right)\right)\theta_n[\alpha, \beta](z|B).$$

Exercise 7.15 Show that the theta functions of order $4n$ with characteristics $[\alpha, \beta]$ span a vector space of dimension n^g, and for a basis of this space one can take the functions

$$\theta\left[\frac{\alpha + \gamma}{n}, \beta\right](nz|nB).$$

Meromorphic functions on an Abelian torus are called *Abelian functions*. An example of an Abelian function is the ratio of two theta functions of the same order with the same θ-characteristics.

7.5 Abel's Theorem

Consider a Riemann surface P with canonical basis of cycles $\{a_i, b_i \mid i = 1, \ldots, g\}$ and the corresponding basis $\{\omega_i\}$ of the space of holomorphic differentials. Set $B_{ik} = \oint_{b_k} \omega_i$.

Lemma 7.7 With a divisor $D = \sum_{i=1}^{n} p_i - \sum_{i=1}^{n} q_i$ on P we associate the differential $\omega = \sum_{i=1}^{n} \tilde{\omega}_i + \sum_{j=1}^{g} n_j \omega_j$ where $\tilde{\omega}_i$ is a meromorphic differential that is holomorphic outside the poles p_i, q_i, has residues $1, -1$ at the points p_i, q_i, respectively, and satisfies the condition $\oint_{a_j} \tilde{\omega}_i = 0$ for all j. Then

$$\oint_{b_k} \omega = \sum_{i=1}^{n} \int_{q_i}^{p_i} \omega_k + \sum_{j=1}^{g} n_j B_{jk}.$$

Proof Applying Riemann's bilinear relation (Theorem 6.9)

$$\sum_{j=1}^{g} (A_j B'_j - A'_j B_j) = 2\pi i \sum_{p} \left(\int_{p_0}^{p} \omega \right) \text{Res}_p(\omega')$$

to the pair of differentials $(\omega_k, \tilde{\omega}_i)$, we obtain

$$2\pi i \oint_{b_k} \tilde{\omega}_i = 2\pi i \int_{q_i}^{p_i} \omega_k.$$

Thus,

$$\oint_{b_k} \omega = \sum_{i=1}^{n} \oint_{b_k} \tilde{\omega}_i + \sum_{j=1}^{g} n_j B_{jk} = \sum_{i=1}^{n} \int_{q_i}^{p_i} \omega_k + \sum_{j=1}^{g} n_j B_{jk}.$$

By Theorem 6.7, the period matrix $B = \{B_{ik}\}$ is a Riemann matrix. The Abelian torus generated by B is called the *Jacobian* $J(P)$ *of the Riemann surface* P.

Exercise 7.16 Show that the Jacobian $J(P)$ does not depend on the choice of a canonical basis of cycles on P.

One can prove that the period matrix uniquely determines the Riemann surface (Torelli's theorem [6, Sec. 2.7]). For $g = 1, 2, 3$, every Riemann matrix is a Jacobian of some Riemann surface. For $g > 3$, according to Theorem 5.7, the Jacobians form a $(3g - 3)$-dimensional subset in the $\frac{g(g+1)}{2}$-dimensional (complex) space of all Riemann matrices. The problem of describing this subset is important for applications and is called the *Schottky problem*.

Fix an arbitrary point $p_0 \in P \setminus \{a_i, b_i\}$. Consider the map

$$A = A_{p_0} \colon p \mapsto \left(\int_{p_0}^{p} \omega_1, \ldots, \int_{p_0}^{p} \omega_g \right) \in \mathbb{C}^g$$

on the surface $P \setminus \{a_i, b_i\}$.

Exercise 7.17 Show that the map $\tilde{A} = \tilde{A}_{p_0} = J_B A \colon P \to J(P)$ is well defined on the surface P.

Exercise 7.18 How does the map $\tilde{A}$ change when the basis of cycles $\{a_i, b_i\}$ changes?

The map $\tilde{A} \colon P \to J(P)$ is called the *Abel–Jacobi map*. It can be extended to an arbitrary divisor $D = \sum_i n_i p_i$ by the formula $\tilde{A}(D) = \sum_i n_i \tilde{A}(p_i) \in J(P)$.

Theorem 7.7 (Abel's Theorem) *A divisor* $D = \sum_{i=1}^{n} p_i - \sum_{i=1}^{n} q_i, \; p_i, q_i \in P$, *is principal if and only if* $A(D) \equiv 0$.

Proof Let $D = \sum_{i=1}^{n} p_i - \sum_{i=1}^{n} q_i$ be the divisor of a meromorphic function f. The periods of the meromorphic differential $\omega = d \ln f$ are multiples of $2\pi i$, i.e.,

$$\oint_{a_k} \omega = 2\pi i \, n_k \quad \text{and} \quad \oint_{b_k} \omega = 2\pi i \, m_k \quad \text{where } n_k, m_k \in \mathbb{Z}.$$

On the other hand, the differential ω can be written as a sum

$$\omega = \sum_{i=1}^{n} \tilde{\omega}_i + \sum_{j=1}^{g} c_j \omega_j$$

where $\tilde{\omega}_i$ is a meromorphic differential that is holomorphic outside the poles p_i, q_i, has residues $1, -1$ at the points p_i, q_i, respectively, and satisfies the condition $\oint_{a_j} \tilde{\omega}_i = 0$ for all j. In particular, $\oint_{a_k} \omega = 2\pi i \, c_k$ and $c_k = n_k$. Thus, by Lemma 7.7, we have

$$\oint_{b_k} \omega = \sum_{i=1}^{n} \int_{q_i}^{p_i} \omega_k + \sum_{j=1}^{g} n_j B_{jk}.$$

Therefore, the kth coordinate of the image $A(D)$ is equal to

$$\left(A\left(\sum_{i=1}^{n} p_i\right) - A\left(\sum_{i=1}^{n} q_i\right) \right)_k = \sum_{i=1}^{n} \int_{q_i}^{p_i} \omega_k = \pi i \, m_k - \sum_{j=1}^{g} n_j B_{jk}.$$

Thus, $A(D)$ belongs to the lattice of quasiperiods, and $A(D) \equiv 0$.
Now let $A(D) \equiv 0$. Then

$$\sum_{i=1}^{n} \int_{q_i}^{p_i} \omega_k = 2\pi i \, m_k - \sum_{j=1}^{g} n_j B_{jk}$$

for some integers n_k, m_k. Consider the differential

$$\omega = \sum_{i=1}^{n} \tilde{\omega}_i + \sum_{j=1}^{g} n_j \omega_j.$$

By Lemma 7.7, we have

$$\oint_{b_k} \omega = \sum_{i=1}^{n} \int_{q_i}^{p_i} \omega_k + \sum_{j=1}^{g} n_j B_{jk} = 2\pi i \, m_k.$$

Hence the function $\exp\left(\int_{p_0}^{p} \omega\right)$ is well defined on P. Moreover, its divisor coincides with D.

7.6 Jacobi Inversion Problem

The Jacobi inversion problem consists in describing the positive divisor $D = \sum_{i=1}^{g} p_i$ mapped by the Abel–Jacobi map A_{p_0} to a prescribed point $\tilde{\xi} \in J(P)$ of the Jacobian.

To solve the Jacobi inversion problem, we will need the function

$$F(p) = F_e(p) = \theta(A(p) - e).$$

In this formula, $\theta(z) = \theta(z|B)$ is the theta function of the surface P corresponding

to the basis $\{a_i, b_i\}$, $e \in \mathbb{C}^g$, and $A(p) = A_{p_0}(p) = \begin{pmatrix} \int_{p_0}^{p} \omega_1 \\ \dots \\ \int_{p_0}^{p} \omega_g \end{pmatrix}$, where the integrals

are taken over paths that do not intersect the cycles $\{a_i, b_i\}$.

Exercise 7.19 Show that the divisor of zeros of the function F_e depends only on the homology classes of the cycles $\{a_i, b_i\}$.

Lemma 7.8 *The set of zeros of the function $F(p)$ either coincides with the whole surface, or is a divisor of degree g.*

Proof Cutting the surface P along the cycles $\{a_i, b_i\}$, we obtain a $4g$-gon $\Gamma \subset \mathbb{C}$ with sides which will be denoted by the same symbols as the corresponding cycles: $a_1, b_1, a_1^{-1}, b_1^{-1}, a_2, b_2, a_2^{-1}, b_2^{-1}, \dots, a_g, b_g, a_g^{-1}, b_g^{-1}$. Denote by A^+ and A^- the

restrictions of the vector functions $\begin{pmatrix} \int_{p_0}^{p} \omega_1 \\ \dots \\ \int_{p_0}^{p} \omega_g \end{pmatrix}$ of p to the sets $\{a_1, b_1, \dots, a_g, b_g\}$ and

$\{a_1^{-1}, b_1^{-1}, \dots, a_g^{-1}, b_g^{-1}\}$, respectively.

Put

$$F^+ = F|_{\{a_1, b_1, \dots, a_g, b_g\}} \quad \text{and} \quad F^- = F|_{\{a_1^{-1}, b_1^{-1}, \dots, a_g^{-1}, b_g^{-1}\}}.$$

To a point $p \in b_k \subset P$ there correspond two points of the boundary. Their images under the Abel–Jacobi map are related by the formula $A_j^+(p) = A_j^-(p) + 2\pi i \, \delta_{jk}$; therefore,

$$d \ln F^-(p) = d \ln F^+(p).$$

The images of a point $p \in a_k \subset P$ are related by the formula $A_j^-(p) = A_j^+(p) + B_{jk}$; therefore,

$$\ln F^-(p) = \ln \left(\theta(A^-(p) - e)\right) = \ln \left(\theta(A^+(p) - e + B_k)\right)$$

$$= \ln \left(\exp\left(-\frac{1}{2}B_{kk} - A_k^+(p) + e_k\right)\theta(A^+(p) - e)\right)$$

$$= -\frac{1}{2}B_{kk} - A_k^+(p) + e_k + \ln F^+(p),$$

whence

$$d \ln F^-(p) = d \ln F^+(p) - \omega_k.$$

Thus, if F is a nonzero function, then the number of its zeros is equal to

$$\frac{1}{2\pi i} \oint_{\partial \Gamma} d \ln F(p) = \frac{1}{2\pi i} \sum_{k=1}^{g} \left(\oint_{a_k} + \oint_{b_k} \right) [d \ln F^+(p) - d \ln F^-(p)]$$

$$= \frac{1}{2\pi i} \sum_{k=1}^{g} \oint_{a_k} \omega_k = g.$$

With a canonical basis of cycles $\{a_i, b_i\}$ we can associate the *vector of Riemann constants*

$$K = K_{p_0} = \begin{pmatrix} K_1 \\ \dots \\ K_g \end{pmatrix} \in \mathbb{C}^g$$

where

$$K_j = \frac{2\pi i + B_{jj}}{2} - \frac{1}{2\pi i} \sum_{\substack{1 \le l \le g, \\ l \ne j}} \oint_{a_l} \omega_l(p) A_j^+(p).$$

Theorem 7.8 (Riemann Vanishing Theorem) *Let* $p_1, \dots, p_g$ *be the set of all zeros of the function* F_e. *Then*

$$A\left(\sum_{i=1}^{g} p_i \right) \equiv e - K$$

where K is the vector of Riemann constants.

Proof The integral

$$\xi_j = \frac{1}{2\pi i} \oint_{\partial \Gamma} (A_j(p)) \, d \ln F(p)$$

is equal to the sum of the residues of the integrand, i.e.,

$$\xi_j = A_j\left(\sum_{i=1}^{g} p_i \right).$$

On the other hand,

$$\xi_j = \frac{1}{2\pi i} \sum_{k=1}^{g} \left(\oint_{a_k} + \oint_{b_k} \right) (A_j^+ \, d \ln F^+ - A_j^- \, d \ln F^-)$$

$$= \frac{1}{2\pi i} \sum_{k=1}^{g} \oint_{a_k} [A_j^+ \, d \ln F^+ - (A_j^+ + B_{jk})(d \ln F^+ - \omega_k)]$$

$$+ \frac{1}{2\pi i} \sum_{k=1}^{g} \oint_{b_k} [A_j^+ \, d \ln F^+ - (A_j^+ - 2\pi i \, \delta_{jk}) \, d \ln F^+]$$

$$= \frac{1}{2\pi i} \sum_{k=1}^{g} \left(\oint_{a_k} A_j^+ \omega_k - B_{jk} \oint_{a_k} d \ln F^+ + 2\pi i \, B_{jk} \right) + \oint_{b_j} d \ln F^+.$$

By the definition of $F^\pm$, we have

$$\oint_{a_k} d \ln F^+ = 2\pi i \, n_k \quad \text{where } n_k \in \mathbb{Z},$$

and

$$\oint_{b_j} d \ln F^+ = \ln F^+(q^+) - \ln F^+(q^-) + 2\pi i \, m$$

$$= \ln \theta(A(q^-) + B_j - e) - \ln \theta(A(q^-) - e) + 2\pi i \, m$$

$$= -\frac{1}{2} B_{jj} + e_j - A_j(q^-) + 2\pi i \, m$$

where $[q^-, q^+]$ is the segment corresponding to the contour b_k and $m \in \mathbb{Z}$.
Besides,

$$A_j \omega_j = \frac{1}{2} d(A_j^2),$$

and hence

$$\oint_{a_j} A_j \omega_k = \frac{1}{2} (A_j^2(p^-) - A_j^2(\tilde{q}))$$

$$= \frac{1}{2} (A_j^2(q^-) - (A_j(q^-) - 2\pi i)^2) = \frac{1}{2} (4\pi i \, A_j(q^-) - (2\pi i)^2)$$

where $[\tilde{q}, q^-]$ is the segment corresponding to the contour a_k^+. Therefore,

$$\frac{1}{2\pi i} \sum_{k=1}^{g} \oint_{a_k} A_j^+ \omega_k = \frac{1}{2\pi i} \sum_{\substack{1 \le k \le g, \\ k \ne j}} \oint_{a_k} A_j^+ \omega_k + A_j(q^-) - \pi i.$$

Thus,

$$\xi_j = e_j - \frac{1}{2} B_{jj} + \frac{1}{2\pi i} \sum_{\substack{1 \le k \le g, \\ k \ne j}} \oint_{a_k} A_j^+ \omega_k - \pi i.$$

The Riemann vanishing theorem implies that $A(S^g P)$ coincides with the Jacobian of the surface P. Let us show that almost every point is the image of only one point of the symmetric power $S^g P$.

By the Riemann–Roch theorem, we have $r(-D) \ge 1 + \deg D - g$, and hence for every positive divisor D of degree greater than g there exists a meromorphic function for which D is the divisor of poles. A positive divisor D of degree g is said to be *special* if there exists a meromorphic function for which D is the divisor of poles, and *nonspecial* otherwise. For instance, the divisor gp is special if and only if p is a Weierstrass point.

Lemma 7.9 *The Abel–Jacobi map $\tilde{A}: S^g P \to J(P)$ from a symmetric power of a Riemann surface is invertible in a neighborhood of a nonspecial divisor.*

Proof If D and D' are positive divisors and $A(D') \equiv A(D)$, then, by Abel's theorem, $D' = D + (f)$. Thus, $D = D'$ if $D = \bigcup p_k$ is a nonspecial divisor. Consider local charts z_k and basic differentials $\omega_i = \varphi_{ik}(z_k) \, dz_k$ in neighborhoods of the points $p_1, \ldots, p_g$. The Jacobian of Abel–Jacobi map at the point D coincides

with the determinant of the matrix $\begin{pmatrix} \varphi_{11} & \cdots & \varphi_{1g} \\ & \cdots & \\ \varphi_{g1} & \cdots & \varphi_{gg} \end{pmatrix}$. This determinant does not

vanish, since otherwise the rows of the matrix would be linearly dependent, which would imply that $i(D) > 0$ and, by the Riemann–Roch theorem, $r(-D) > 1$. Therefore, the Jacobian of the Abel–Jacobi map does not vanish in a neighborhood of the divisor D, too.

One can also prove (see, e.g., [6, Vol. 1]) the following theorem due to Riemann.

Theorem 7.9 *The vector of Riemann constants K satisfies the following relations: $2K \equiv -A(\mathcal{K})$, where $\mathcal{K}$ is the canonical class, and*

$$\{\zeta \in \mathbb{C}^g \mid \theta(\zeta) = 0\} \equiv A(S^{g-1} P) + K.$$

The function $F_e(p) = \theta(A_{p_0}(p) - e)$ is identically zero if and only if $e \equiv K + A(D)$ where D is a special divisor.

Chapter 8
Integrable Systems

8.1 Formal Exponentials

In what follows, we use the binomial coefficients $\binom{p}{t}$ given by $\binom{p}{t} = \frac{p!}{t!(p-t)!}$ for $p \geq t \geq 0$ and $\binom{p}{t} = 0$ otherwise, with the standard convention that $0! = 1$.

Set $\partial_n = \frac{\partial}{\partial x_n}$ and $\partial = \partial_1$.

Definition 8.1 By a *formal exponential* in a local parameter $z \in \mathbb{C}$ in a neighborhood of ∞ we mean a formal function in infinitely many variables $x = (x_1, x_2, \ldots)$ of the form

$$\psi(z, x) = \exp\left(\sum_{i=1}^{\infty} x_i z^i\right)\left(1 + \sum_{j=1}^{\infty} \xi_j(x) z^{-j}\right)$$

for which there exist operators

$$L_n = \partial^n + \sum_{i=2}^{n} B_n^i(x) \partial^{n-i}, \quad n = 1, 2, \ldots,$$

such that $\partial_n \psi = L_n \psi$.

Exercise 8.1 Show that in this case

$$B_s^t = -\sum_{i=1}^{t-1} \binom{s}{i} \partial^i \xi_{t-i} - \sum_{j=2}^{t-1} B_s^j \sum_{i=0}^{t-j-1} \binom{s-j}{i} \partial^i \xi_{t-i-j},$$

$$\partial_n \xi_i = \sum_{j=1}^{n+i-1} \binom{n}{j} \partial^j \xi_{i+n-j} + \sum_{k=2}^{n} B_n^k \sum_{j=0}^{n-k} \binom{n-k}{j} \partial^j \xi_{i+n-j-k}.$$

© Springer Nature Switzerland AG 2019
S. M. Natanzon, *Complex Analysis, Riemann Surfaces and Integrable Systems*,
Moscow Lectures 3, https://doi.org/10.1007/978-3-030-34640-9_8

The coefficients $\eta_j(x)$ of the expansion

$$\ln \psi(x, z) = \sum_{j=1}^{\infty} x_j z^j + \sum_{j=1}^{\infty} \eta_j z^{-j}$$

are related to the functions $\xi_j(x)$ by the formula

$$\xi_j = \sum_{n=1}^{\infty} \frac{1}{n!} \sum_{i_1+\ldots+i_n=j} \eta_{i_1} \ldots \eta_{i_n}.$$

This allows one to express the functions $B_s^t(x)$ in terms of η_j. To describe this dependence, we will need combinatorial constants $P_s \begin{pmatrix} i_1 & \cdots & i_n \\ j_1 & \cdots & j_n \end{pmatrix}$ that depend on positive integers $s, i_1, \ldots, i_n$ and nonnegative integers $j_1, \ldots, j_n$. They are defined by the recurrence relation

1. $P_s \begin{pmatrix} i_1 & \cdots & i_n \\ 0 & \cdots & 0 \end{pmatrix} = 0;$

2. $P_s \begin{pmatrix} i \\ j \end{pmatrix} = \binom{s}{j}$ for $j > 0;$

3. $P_s \begin{pmatrix} i_1 & \cdots & i_n \\ j_1 & \cdots & j_n \end{pmatrix} = \frac{1}{n!} \binom{s}{j_1+\ldots+j_n} \frac{(j_1+\ldots+j_n)!}{j_1!\ldots j_n!}$

$$- \sum_{q=1}^{n-1} P_s \begin{pmatrix} i_1 & \cdots & i_q \\ j_1 & \cdots & j_q \end{pmatrix} \frac{1}{(n-q)!} \binom{s-(i_1+\ldots+i_q+j_1+\ldots+j_q)}{j_{q+1}+\ldots+j_n} \frac{(j_{q+1}+\ldots+j_n)!}{j_{q+1}!\ldots j_n!}$$

for $(j_1, \ldots, j_n) \neq (0, \ldots, 0)$.

Denote by $\begin{bmatrix} i_1 & \cdots & i_n \\ j_1 & \cdots & j_n \end{bmatrix}$ the set of all matrices that can be obtained from the matrix $\begin{pmatrix} i_1 & \cdots & i_n \\ j_1 & \cdots & j_n \end{pmatrix}$ by permuting columns. Let $\left\| \begin{matrix} i_1 & \cdots & i_n \\ j_1 & \cdots & j_n \end{matrix} \right\|$ be the number of such matrices.

Put $P_s \begin{bmatrix} i_1 & \cdots & i_n \\ j_1 & \cdots & j_n \end{bmatrix} = \sum P_s \begin{pmatrix} a_1 & \cdots & a_n \\ b_1 & \cdots & b_n \end{pmatrix}$ where the sum is over all matrices $\begin{pmatrix} a_1 & \cdots & a_n \\ b_1 & \cdots & b_n \end{pmatrix}$ from the set $\begin{bmatrix} i_1 & \cdots & i_n \\ j_1 & \cdots & j_n \end{bmatrix}$.

The following lemma can be proved by induction.

Lemma 8.1 Let $m > 0, k > 0$, and $j_t \geq 1$ for $t \leq m$. Then

$$P_s \begin{bmatrix} i_1 & \cdots & i_m & i_{m+1} & \cdots & i_{m+k} \\ j_1 & \cdots & j_m & 0 & \cdots & 0 \end{bmatrix} = 0$$

if $s \geq i_1 + \ldots + i_m + j_1 + \ldots + j_m$, *and*

$$P_s \begin{bmatrix} i_1 & \ldots & i_m & i_{m+1} & \ldots & i_{m+k} \\ j_1 & \ldots & j_m & 0 & \ldots & 0 \end{bmatrix} = \frac{1}{k!} \left\| \begin{matrix} i_{m+1} & \ldots & i_{m+k} \\ 0 & \ldots & 0 \end{matrix} \right\| P_s \begin{bmatrix} i_1 & \ldots & i_m \\ j_1 & \ldots & j_m \end{bmatrix}$$

if $s < i_1 + \ldots + i_m + j_1 + \ldots + j_m$.

Comparing Exercise 8.1 and Lemma 8.1, we obtain the following result.

Lemma 8.2 *Let* $2 \leq t \leq s$. *Then*

$$B_s^t = -\sum_{n=1}^{\infty} \sum P_s \begin{pmatrix} i_1 & \ldots & i_n \\ j_1 & \ldots & j_n \end{pmatrix} \partial^{j_1} \eta_{i_1} \ldots \partial^{i_n} \eta_{j_n}$$

where the second sum is over all matrices $\begin{pmatrix} i_1 & \ldots & i_n \\ j_1 & \ldots & j_n \end{pmatrix}$ *such that* $i_m, j_m \geq 1$ *and*
$i_1 + \ldots + i_n + j_1 + \ldots + j_n = t$.

Proof We use induction on t. For $t = 2$, by Exercise 8.1, we have

$$B_s^2 = -s \, \partial \xi_1 = -P_s \begin{pmatrix} 1 \\ 1 \end{pmatrix} \partial \eta_1.$$

Let us prove the assertion for $t = N$ assuming that it holds for $t < N$. By
Exercise 8.1, we have

$$B_s^t = -\sum_{i=1}^{t-1} \binom{s}{i} \partial^i \left(\sum_{n=1}^{\infty} \frac{1}{n!} \sum_{i_1+\ldots+i_n=t-i} \eta_{i_1} \ldots \eta_{i_n} \right)$$

$$+ \sum_{j=2}^{t-1} \left(\sum_{n=1}^{\infty} \sum_{i_1+\ldots+j_n=j} P_s \begin{pmatrix} i_1 & \ldots & i_n \\ j_1 & \ldots & j_n \end{pmatrix} \partial^{j_1} \eta_{i_1} \ldots \partial^{j_n} \eta_{i_n} \right)$$

$$\times \left(\sum_{i=0}^{t-j-1} \binom{s-j}{i} \partial^j \left(\sum_{n=1}^{\infty} \frac{1}{n!} \sum_{i_1+\ldots+i_n=t-i-j} \eta_{i_1} \ldots \eta_{i_n} \right) \right)$$

$$= -\sum_{n=1}^{\infty} \sum \left(\frac{1}{n!} \binom{s}{j_1 + \ldots + j_n} \frac{(j_1 + \ldots + j_n)!}{j_1! \ldots j_n!} - \sum_{q=1}^{n-1} P_s \begin{pmatrix} i_1 & \ldots & i_q \\ j_1 & \ldots & j_q \end{pmatrix} \right.$$

$$\times \frac{1}{(n-q)!} \binom{s - (i_1 + \ldots + i_q + j_1 + \ldots + j_q)}{j_{q+1} + \ldots + j_n} \frac{(j_{q+1} + \ldots + j_n)!}{j_{q+1}! \ldots j_n!} \right) \partial^{j_1} \eta_{i_1} \ldots \partial^{j_n} \eta_{i_n}$$

$$= -\sum_{n=1}^{\infty} \sum P_s \begin{pmatrix} i_1 & \ldots & i_n \\ j_1 & \ldots & j_n \end{pmatrix} \partial^{j_1} \eta_{i_1} \ldots \partial^{j_n} \eta_{i_n}$$

where the second sum is over all matrices $\begin{pmatrix} i_1 & \ldots & i_n \\ j_1 & \ldots & j_n \end{pmatrix}$ such that $i_m \geq 1$, $j_m \geq 0$,
and $i_1 + \ldots + i_n + j_1 + \ldots + j_n = t$. By Lemma 8.1, we may assume that the last
sum is taken over only positive $j_m > 0$.

Theorem 8.1 *For all $r \geq 1$ and $s \geq 1$,*

$$\partial_s \eta_r = \sum_{n=1}^{\infty} \sum P_s \begin{pmatrix} i_1 & \cdots & i_n \\ j_1 & \cdots & j_n \end{pmatrix} \partial^{j_1} \eta_{i_1} \cdots \partial^{j_n} \eta_{i_n}$$

where the second sum is over all matrices $\begin{pmatrix} i_1 & \cdots & i_n \\ j_1 & \cdots & j_n \end{pmatrix}$ such that $i_m \geq 1$, $j_m \geq 1$, and $i_1 + \ldots + i_n + j_1 + \ldots + j_n = r + s$.

Proof We use induction on r. For $r = 1$, by Exercise 8.1 and Lemma 8.2, we have

$$\partial_s \eta_1 = \partial_s \xi_1 = \sum_{j=1}^{s} \binom{s}{j} \partial^j \xi_{s+1-j} + \sum_{k=2}^{s} B_s^k \sum_{j=0}^{s-k} \binom{s-k}{j} \partial^j \xi_{1+s-j-k}$$

$$= \sum_{j=1}^{s} \binom{s}{j} \partial^j \xi_{s+1-j} - \sum_{k=2}^{\infty} \left(\sum_{n=1}^{\infty} \sum_{i_1+\ldots+j_n=k} P_s \begin{pmatrix} i_1 & \cdots & i_n \\ j_1 & \cdots & j_n \end{pmatrix} \partial^{j_1} \eta_{i_1} \cdots \partial^{j_n} \eta_{i_n} \right)$$

$$\times \left(\sum_{j=0}^{s-k} \binom{s-k}{j} \partial^j \xi_{1+s-j-k} \right) = \sum_{n=1}^{\infty} \sum P_s \begin{pmatrix} i_1 & \cdots & i_n \\ j_1 & \cdots & j_n \end{pmatrix} \partial^{j_1} \eta_{i_1} \cdots \partial^{j_n} \eta_{i_n}$$

where the second sum is over all matrices $\begin{pmatrix} i_1 & \cdots & i_n \\ j_1 & \cdots & j_n \end{pmatrix}$ such that $i_m \geq 1$, $j_m \geq 0$, and $i_1 + \ldots + i_n + j_1 + \ldots + j_n = s + 1$. By Lemma 8.1, we may assume that the sum is taken over only positive $j_m > 0$.

Thus, for $r = 1$ the theorem is proved.

Now let us prove the assertion for $r = N$ assuming that it holds for $r < N$. By Exercise 8.1, we have

$$\partial_s \left(\sum_{n=1}^{\infty} \frac{1}{n!} \sum_{i_1+\ldots+i_n=r} \eta_{i_1} \cdots \eta_{i_n} \right)$$

$$= \sum_{j=1}^{s+r-1} \binom{s}{j} \partial^j \xi_{s+r-j} + \sum_{k=2}^{\infty} B_s^k \sum_{j=0}^{s-k} \binom{s-k}{j} \partial^j \xi_{r+s-j-k}.$$

Thus, by Lemma 8.2, Exercise 8.1, and the induction hypothesis, we have

$$\partial_s \eta_r = \sum_{j=1}^{\infty} \binom{s}{j} \partial^j \left(\sum_{n=1}^{\infty} \frac{1}{n!} \sum_{i_1+\ldots+i_n=s+r-j} \eta_{i_1} \cdots \eta_{i_n} \right)$$

$$+ \sum_{k=2}^{\infty} \left(\sum_{i_1+\ldots+j_n=k} P_s \begin{pmatrix} i_1 & \cdots & i_n \\ j_1 & \cdots & j_n \end{pmatrix} \partial^{j_1} \eta_{i_1} \cdots \partial^{j_n} \eta_{i_n} \right)$$

$$\times \sum_{j=0}^{s-k} \binom{s-k}{j} \partial^j \Big(\sum_{n=1}^{\infty} \frac{1}{n!} \sum_{i_1+\ldots+i_n=r+s-j-k} \eta_{i_1}\cdots\eta_{i_n}\Big)$$

$$-\partial_s \Big(\sum_{n=2}^{\infty} \frac{1}{n!} \sum_{i_1+\ldots+i_n=r} \eta_{i_1}\cdots\eta_{i_n}\Big)$$

$$= \sum_{n=1}^{\infty} \sum P_s\begin{pmatrix} i_1 & \cdots & i_n \\ j_1 & \cdots & j_n \end{pmatrix} \partial^{j_1}\eta_{i_1}\cdots\partial^{j_n}\eta_{i_n}$$

where the second sum is over all matrices $\begin{pmatrix} i_1 & \cdots & i_n \\ j_1 & \cdots & j_n \end{pmatrix}$ such that $i_m \geq 1$, $j_m \geq 1$, and $i_1 + \ldots + i_n + j_1 + \ldots + j_n = s + r$.

8.2   The KP Hierarchy

Definition 8.2 A formal function $\tau(x) = \tau(x_1, x_2, \ldots)$ will be called an *almost tau function of the KP hierarchy* if the formal function

$$\psi(z, x) = \exp\Big(\sum x_j\Big(z^j + \sum_{i=1}^{\infty} a_{ij}z^{-i}\Big)\Big) \frac{\tau(x_1 - z^{-1}, x_2 - \frac{1}{2}z^{-2}, x_3 - \frac{1}{3}z^{-3}, \ldots)}{\tau(x_1, x_2, x_3, \ldots)}$$

is a formal exponential in the local parameter $z \in \mathbb{C}$ in a neighborhood of ∞ for some collection of constants a_{ij}. If all constants a_{ij} are zero, then ψ is called a *tau function*.

Our definition of a tau function agrees with those from [1] and [4].

Let us represent the system of differential equations from Lemma 8.2 as a system of differential equations for the function

$$v(x) = \ln \tau(x).$$

It is not difficult to see that

$$\sum_{j=1}^{\infty} \eta_j z^{-j} = \ln \psi(x, z) - \sum_{j=1}^{\infty} x_j z^j$$

$$= \sum_{i,j=1}^{\infty} a_{ij}x_j z^{-i} + \ln \tau(x_1 - z^{-1}, x_2 - \frac{1}{2}z^{-2}, \ldots) - \ln \tau(x_1, x_2, \ldots)$$

$$= \sum_{i,j=1}^{\infty} a_{ji} x_i z^{-j} + v(x_1 - z^{-1}, x_2 - \frac{1}{2} z^{-2}, \dots) - v(x)$$

$$= \sum_{i,j=1}^{\infty} a_{ji} x_i z^{-j} + \sum_{n=1}^{\infty} \sum_{i_1 + \dots + i_n = j} \frac{(-1)^n}{n! i_1 \dots i_n} \partial_{i_1} \dots \partial_{i_n} v(x) z^{-j}.$$

Thus,

$$\eta_r = \sum_{n=1}^{\infty} \sum_{i_1 + \dots + i_n = r} \frac{(-1)^n}{n! i_1 \dots i_n} \partial_{i_1} \dots \partial_{i_n} v + \sum_{i=1}^{\infty} a_{ri} x_i. \tag{8.1}$$

Theorem 8.2 *Assume that all constants a_{ri} are zero. Then there exist universal rational coefficients*

$$R_r \begin{pmatrix} s_1 & \dots & s_n \\ t_1 & \dots & t_n \end{pmatrix}, \quad R_{ij} \begin{pmatrix} s_1 & \dots & s_n \\ t_1 & \dots & t_n \end{pmatrix}$$

such that

$$\eta_r = -\frac{1}{r} \partial_r v + \sum_{n=1}^{\infty} \sum \sum R_r \begin{pmatrix} s_1 & \dots & s_n \\ t_1 & \dots & t_n \end{pmatrix} \partial_{s_1} \partial^{t_1} v \dots \partial_{s_n} \partial^{t_n} v, \tag{8.2}$$

$$\partial_i \partial_j v = \sum_{n=1}^{\infty} \sum \sum R_{ij} \begin{pmatrix} s_1 & \dots & s_n \\ t_1 & \dots & t_n \end{pmatrix} \partial_{s_1} \partial^{t_1} v \dots \partial_{s_n} \partial^{t_n} v \tag{8.3}$$

where the last sum is over all matrices $\begin{pmatrix} s_1 & \dots & s_n \\ t_1 & \dots & t_n \end{pmatrix}$ such that $s_m, t_m \geq 1$ and the sum $s_1 + \dots + s_n + t_1 + \dots + t_n$ is equal to r in (8.2) and to $i + j$ in (8.3).

Proof We prove the theorem by a simultaneous induction on k and $i + j$. For $i + j = 2$, it is obvious. For $r = 1$, it follows from (8.1). Let us prove the theorem for $i + j = N$ and $r = N - 1$ assuming that it holds for $i + j < N$ and $r < N - 1$. Below we assume that $s_m, t_m \geq 1$ and $\sigma_n = s_1 + \dots + s_n + t_1 + \dots + t_n$. Then, in view of (8.1) and (8.3),

$$\eta_r = -\frac{1}{r} \partial_r v + \sum_{n=2}^{\infty} \sum_{s_1 + \dots + s_n = r} \frac{(-1)^n}{n! s_1 \dots s_n} \partial_{s_1} \dots \partial_{s_n} v$$

$$= -\frac{1}{r} \partial_r v + \sum_{n=1}^{\infty} \sum_{\sigma_n = r} R_r \begin{pmatrix} s_1 & \dots & s_n \\ t_1 & \dots & t_n \end{pmatrix} \partial_{s_1} \partial^{t_1} v \dots \partial_{s_n} \partial^{t_n} v.$$

Thus, by (8.2), (8.3), and Theorem 8.1, we have

$$-\frac{1}{j}\partial_i\partial_j v = \partial_i\eta_j - \partial_i\left(\sum_{n=1}^{\infty}\sum_{\sigma_n=j} R_j\begin{pmatrix} s_1 \cdots s_n \\ t_1 \cdots t_n \end{pmatrix}\partial_{s_1}\partial^{t_1}v\ldots\partial_{s_n}\partial^{t_n}v\right)$$

$$=\sum_{n=1}^{\infty}\sum_{\sigma_n=i+j} P_i\begin{pmatrix} s_1 \cdots s_n \\ t_1 \cdots t_n \end{pmatrix}\partial^{t_1}\eta_{s_1}\ldots\partial^{t_n}\eta_{s_n}$$

$$-\partial_i\left(\sum_{n=1}^{\infty}\sum_{\sigma_n=j} R_j\begin{pmatrix} s_1 \cdots s_n \\ t_1 \cdots t_n \end{pmatrix}\partial^{t_1}\partial_{s_1}v\ldots\partial^{t_n}\partial_{s_n}v\right)$$

$$=\sum_{n=1}^{\infty}(-1)^{n+1}\sum_{s_1+\ldots+s_n+n=i+j} P_i\begin{pmatrix} s_1 \cdots s_n \\ 1 \cdots 1 \end{pmatrix}\partial\left(\frac{1}{s_1}\partial_{s_1}v\right)\ldots\partial\left(\frac{1}{s_n}\partial_{s_n}v\right)$$

$$+\sum_{n=1}^{\infty}\sum_{\sigma_n=i+j,t_1+\ldots+t_n>n} R_{ij}\begin{pmatrix} s_1 \cdots s_n \\ t_1 \cdots t_n \end{pmatrix}\partial_{s_1}\partial^{t_1}v\ldots\partial_{s_n}\partial^{t_n}v.$$

The procedure described in the proof of Theorem 8.2 allows one to explicitly find all rational constants $R_{ij}\begin{pmatrix} s_1 \cdots s_m \\ t_1 \cdots t_m \end{pmatrix}$. Here are several first equations of the hierarchy (8.3):

$$\partial_2^2 v = \frac{4}{3}\partial_3\partial v - \frac{1}{3}\partial^4 v - 2\,(\partial^2 v)^2,$$

$$\partial_3\partial_2 v = \frac{3}{2}\partial_4\partial v - \frac{3}{2}\partial_2\partial^3 v - 3\,\partial_2\partial v\,\partial^2 v,$$

$$\partial_3^2 v = \frac{9}{5}\partial_5\partial v - \partial_3\partial^3 v + \frac{1}{5}\partial^6 v - 3\,\partial_3\partial v\,\partial^2 v - \frac{9}{4}\,(\partial_2\partial v)^2$$

$$+ 3\,\partial^4 v\,\partial^2 v + \frac{9}{4}\,(\partial^3 v)^2 + 3\,(\partial^2 v)^3. \tag{8.4}$$

Theorem 8.2 implies that a formal solution to the *KP hierarchy* (8.3) (i.e., a solution expressed in the form of a power series) is uniquely determined by an infinite collection $f_i(x_1) = \partial_i v|_{x_2=x_3=\ldots=0}$ $(i = 1, 2, \ldots)$ of series in one variable. It is natural to regard these functions as Cauchy data for the hierarchy. In the recent paper [22], it was proved that to every collection of such Cauchy data there corresponds a solution to the hierarchy, and an algorithm was suggested for constructing this solution.

In applications related to wave processes, one usually uses the function $u(x) = 2\partial^2 \ln \tau(x) = 2\partial^2 v$. Equation (8.4) implies the following result.

Corollary 8.1 *Let* $\tau(x)$ *be a formal tau function. Then the function* $u(x) = 2\partial^2 \ln \tau(x)$ *satisfies the Kadomtsev–Petviashvili equation*

$$\partial_2^2 u = \frac{4}{3} \partial_3 \partial u - \frac{1}{3} \partial^4 u - 2 \left((\partial u)^2 + u \partial^2 u \right).$$

8.3   The *n*-KdV Hierarchy

The system of differential equations (8.3) together with the additional equation $\partial_n v = 0$ is called the *n-KdV hierarchy*, or the *Gelfand–Dikii hierarchy*. In this case, by Theorem 8.2, we have

$$0 = \partial_m \partial_n v = \frac{mn}{m+n-1} \partial_{n+m-1} \partial v + \sum_{m=1}^{\infty} \sum R_{mn} \begin{pmatrix} s_1 & \cdots & s_m \\ t_1 & \cdots & t_m \end{pmatrix} \partial_{s_1} \partial^{t_1} v \ldots \partial_{s_m} \partial^{t_m} v$$

where $1 \leq s_j \leq n+m-2$, $t_j \geq 1$. This allows one to recursively express the functions $\partial_k \partial v$ for $k > n$ in terms of the functions $\partial_r \partial v$ with $r < n$, i.e., to find a relation

$$\partial_{n+r} \partial v = \sum_{m=1}^{\infty} \sum N_{(n+r)1}^m \begin{pmatrix} s_1 & \cdots & s_m \\ t_1 & \cdots & t_m \end{pmatrix} \partial_{s_1} \partial^{t_1} v \ldots \partial_{s_m} \partial^{t_m} v \qquad (8.5)$$

where $t_j \geq 1$, $s_j < n$, $\sum_{j=1}^{m} (s_j + t_j) = n + r + 1$.

For $n = 2$, the system (8.5) is called the *KdV hierarchy*.

Comparing the systems (8.5) and (8.3), we find the *n*-KdV hierarchy

$$\partial_i \partial_j v = \sum_{m=1}^{\infty} \sum N_{ij}^m \begin{pmatrix} s_1 & \cdots & s_m \\ t_1 & \cdots & t_m \end{pmatrix} \partial_{s_1} \partial^{t_1} v \ldots \partial_{s_m} \partial^{t_m} v \qquad (8.6)$$

where $i, j \geq 1$, $1 \leq s_\alpha \leq n - 1$, $t_\alpha \geq 1$, $\sum_{\alpha=1}^{m} (s_\alpha + t_\alpha) = i + j$.

The coefficients $N_{i,j}^m \begin{pmatrix} s_1 & \cdots & s_m \\ t_1 & \cdots & t_m \end{pmatrix}$ are rational constants. The constructions involved in the definition of these constants lead to recurrence relations allowing one to compute them.

The structure of (8.6) ensures that a formal solution to this system is uniquely, up to a constant, determined by an arbitrary collection $f_i(x_1) = \partial_i v|_{x_2=x_3=\ldots=0}$ ($i = 1, \ldots, n-1$) of $n-1$ series in one variable, which can be interpreted as Cauchy data for the *n*-KdV hierarchy. Moreover, Eq. (8.5) allow one to find the

Cauchy data for the corresponding solution to the KP hierarchy. This, in turn, allows one to determine a solution to the n-KdV hierarchy from its Cauchy data.

As before, the function $u(x) = 2\partial^2 \ln \tau(x) = 2\partial^2 v$ gives solutions to equations important for applications.

Examples

1. For $n = 2$, the first equation in (8.6) is the *KdV equation*

$$\partial_3 u = \frac{3}{2} u \, \partial u + \frac{1}{4} \partial^3 u.$$

2. For $n = 3$, the first equation in (8.6) is the *Boussinesq equation*

$$\partial_2^2 u = -\frac{1}{3} \partial^4 u - \partial^2(u^2).$$

8.4 Baker–Akhiezer Functions

A Baker–Akhiezer function $\psi \colon P \to \bar{\mathbb{C}}$ is an analog of the exponential of a polynomial for a Riemann surface P of arbitrary genus g. It depends on

- a point $p_0 \in P$ (which will be called the *special* point);
- a local chart $z \colon U \to \bar{\mathbb{C}}$ in a neighborhood of p_0 for which $z(p_0) = \infty$;
- a positive nonspecial divisor $D \in P \setminus p_0$ of degree g;
- a generic polynomial $q(z) = q_1 z + \ldots + q_n z^n$ of positive degree.

A *Baker–Akhiezer function* is a function $\psi \colon P \to \bar{\mathbb{C}}$ satisfying the following properties:

- the function ψ is meromorphic on $P \setminus p_0$, and its divisor of poles coincides with D;
- the function $\psi \exp[-q(z(p))]$ is analytic and has no zeros in a neighborhood of p_0.

The set of all Baker–Akhiezer functions with parameters (p_0, z, D, q) is a vector space $BA(p_0, z, D, q)$.

Choose a canonical basis of cycles $\{a_i, b_i\}$ on P. Denote by Ω^q a meromorphic differential that has a unique pole at the point p_0, has the form

$$\Omega^q(p) = dq(z) + O(1) \, d\varepsilon = q' \, dz + O(z^{-2}) \, dz$$

in the local chart $\varepsilon = \frac{1}{z}$ in a neighborhood of p_0, and satisfies the conditions

$$\oint_{a_i} \Omega^q = 0 \quad (i = 1, \ldots, g).$$

Let U^q be the vector with components

$$U_i^q = \oint_{b_i} \Omega^q.$$

Lemma 8.3 *The degree of the divisor of zeros D' of a Baker–Akhiezer function ψ is equal to g, and $A(D') \equiv A(D) - U^q$.*

Proof Consider the meromorphic differential $\omega = d \ln \psi = \frac{d\psi}{\psi}$. The function

$$\ln(\psi \exp[-q(z(p))]) = \ln(\psi) - q(z(p))$$

is analytic in a neighborhood of p_0. Therefore, in a neighborhood of p_0 the differentials ω and Ω^q have the same principal parts and, in particular, zero residues. Thus, the poles of ω form the divisor $D + D'$. Moreover, its residues are equal to 1 at the points of the divisor D' and -1 at the points of the divisor D. Thus, by the theorem on the sum of the residues of a differential, $\deg D' = \deg D = g$.

Let $D = \sum_{j=1}^{g} p_j$ and $D' = \sum_{j=1}^{g} p'_j$. Denote by $\tilde{\omega}_j$ a meromorphic differential that is holomorphic outside the points p_j, p'_j, has residues -1, 1 at p_j, p'_j, respectively, and is normalized so that $\oint_{a_i} \tilde{\omega}_j = 0$ $(i = 1, \ldots, g)$. Then

$$\omega = \sum_{j=1}^{g} \tilde{\omega}_j + \Omega^q + \sum_{r=1}^{g} c_r \omega_r$$

where $\{\omega_r\}$ is a basis of holomorphic differentials normalized so that $\oint_{a_i} \omega_r = 2\pi i \, \delta_{ir}$. On the other hand,

$$\oint_{a_i} \omega = 2\pi i \, n_i, \quad \oint_{b_i} \omega = 2\pi i \, m_i \quad \text{where } n_i, m_i \in \mathbb{Z}.$$

Thus, $c_i = n_i$, and, by Riemann's bilinear relations (Theorem 6.9),

$$2\pi i \, m_i = \oint_{b_i} \left(\sum_{j=1}^{g} \tilde{\omega}_j + \Omega^q + \sum_{r=1}^{g} c_r \omega_r \right) = \sum_{j=1}^{g} \int_{p_j}^{p'_j} \tilde{\omega}_j + U_i^q + \sum_{j=1}^{g} c_j B_{jk};$$

therefore,

$$(A(D') - A(D))_i = A_i \left(\sum_{j=1}^{g} \int_{p_j'}^{p_j} \tilde{\omega}_j \right) \equiv -U_i^q.$$

Theorem 8.3 *The vector space $BA(p_0, z, D, q)$ is one-dimensional and is spanned by the function*

$$\psi = \exp \left(\int_{p_*}^{p} \Omega^q \right) \frac{\theta(A_{p_*}(p) - A_{p_*}(D) + U^q - K_{p_*})}{\theta(A_{p_*}(p) - A_{p_*}(D) - K_{p_*})}$$

where $p_* \neq p_0$, K_{p_*} is the vector of Riemann constants, and the integration paths for $A_{p_*}(p)$ and $\int_{p_*}^{p} \Omega^q$ coincide.

Proof The denominator does not vanish identically, since the divisor D is nonspecial (Theorem 7.9). Moreover, the divisor of poles of the function ψ coincides with the divisor of zeros of the function $F_{A_{p_*}(D)+K_{p_*}} = \theta(A_{p_*}(p) - A_{p_*}(D) - K_{p_*})$. By the Riemann vanishing theorem (Theorem 7.8), the image of this divisor under the Abel–Jacobi map is equal to $A_{p_*}(D) + K_{p_*} - K_{p_*} = A_{p_*}(D)$. Therefore, by Lemma 7.9, the divisor of poles of the function ψ coincides with D.

Let us prove that the right-hand side does not depend on the integration path. Indeed, if we extend the integration path by a cycle $c = \sum_{i=1}^{g} n_i a_i + \sum_{i=1}^{g} m_i b_i$, then the integral $\int_{p_*}^{p} \Omega^q$ increases by

$$\sum_{i=1}^{g} m_i U_i^q = \langle M, U^q \rangle \quad \text{where } M = (m_1, \ldots, m_g),$$

and the vector $A_{p_*}(p)$ increases by

$$2\pi i N + BM \quad \text{where } N = (n_1, \ldots, n_g).$$

Therefore, the function $\psi_{(p_0, \varepsilon, D, q)}$ gets multiplied by

$$\exp\langle M, U^q \rangle \frac{\exp\left[-\frac{1}{2} \langle BM, M \rangle - \langle M, A_{p_*}(p) - A_{p_*}(D) + U^q - K_{p_*} \rangle \right]}{\exp\left[-\frac{1}{2} \langle BM, M \rangle - \langle M, A_{p_*}(p) - A_{p_*}(D) - K_{p_*} \rangle \right]} = 1.$$

Thus, $\psi \in BA(p_0, z, D, q)$ and, consequently, $BA(p_0, z, D, q) \neq \varnothing$.

Now consider an arbitrary function $\tilde{\psi} \in BA(p_0, z, D, q)$ with divisor of zeros $\tilde{D}'$. By Lemma 8.3, we have $A(\tilde{D}') = A(D) - U^q = A(D')$. Moreover, the divisors D' and $\tilde{D}'$ are nonspecial, since D is a nonspecial divisor and U^q is a generic vector. Thus, by Lemma 7.9, we have $D' = \tilde{D}'$. Therefore, $\frac{\tilde{\psi}}{\psi} = \text{const}$.

Exercise 8.2 Construct functions on a Riemann surface that behave as exponentials at several points, and prove for them an analog of the last theorem.

8.5   Normalized Baker–Akhiezer Functions

Now let us study the dependence of a Baker–Akhiezer function from the set $BA(p_0, z, D, q)$ on the coefficients of the polynomial $q(z) = \sum_{i=1}^{n} x_i z^i$. For our purposes, it is convenient to consider the "universal polynomial" $q(z, x) = \sum_{i=1}^{\infty} x_i z^i$ where $x = (x_1, x_2, \dots)$ but only finitely many coordinates x_i do not vanish. Every such collection x yields a one-dimensional vector space of Baker–Akhiezer functions $BA(p_0, z, D, x) = BA(p_0, z, D, q(z, x))$.

Denote by Ω^i a meromorphic differential with zero a-periods that has a unique pole at the point p_0 and has the form

$$\Omega^i(p) = d(z^i) + O(z^{-2}) \, dz, \tag{8.7}$$

and denote by U^i the vector with coordinates

$$U_j^i = \oint_{b_j} \Omega^i \quad (j = 1, \dots, g).$$

Then $\Omega^q = \sum_{i=1}^{\infty} x_i \Omega^i$ and $U^q = \sum_{i=1}^{\infty} x_i U^i$.

Lemma 8.4 *The vector space $BA(p_0, z, D, x)$ is spanned by the function*

$$\psi(z, x) = \psi(p_0, p(z), D, x)$$

$$= \exp\left(\int_{p_*}^{p} \sum_{i=1}^{\infty} x_i \Omega^i \right) \frac{\theta(A_{p_*}(p) - A_{p_*}(D) + \sum\limits_{i=1}^{\infty} x_i U^i - K_{p_*}) \theta(A_{p_*}(p_0) - A_{p_*}(D) - K_{p_*})}{\theta(A_{p_*}(p) - A_{p_*}(D) - K_{p_*}) \theta(A_{p_*}(p_0) - A_{p_*}(D) + \sum\limits_{i=1}^{\infty} x_i U^i - K_{p_*})}.$$

In the local chart z, it has the form

$$\psi(z, x) = \exp\left(\sum_{i=1}^{\infty} x_i z^i\right)\left(1 + \sum_{j=1}^{\infty} \xi_i z^{-j}\right).$$

Proof The function

$\psi(z, x)$

$$= \exp\left(\int_{p_*}^{p} \sum_{i=1}^{\infty} x_i \Omega^i\right) \frac{\theta(A_{p_*}(p) - A_{p_*}(D) + \sum\limits_{i=1}^{\infty} x_i U^i - K_{p_*})\theta(A_{p_*}(p_0) - A_{p_*}(D) - K_{p_*})}{\theta(A_{p_*}(p) - A_{p_*}(D) - K_{p_*})\,\theta(A_{p_*}(p_0) - A_{p_*}(D) + \sum\limits_{i=1}^{\infty} x_i U^i - K_{p_*})}$$

differs from the function

$$\exp\left(\int_{p_*}^{p} \sum_{i=1}^{\infty} x_i \Omega^i\right) \frac{\theta(A_{p_*}(p) - A_{p_*}(D) + \sum\limits_{i=1}^{\infty} x_i U^i - K_{p_*})}{\theta(A_{p_*}(p) - A_{p_*}(D) - K_{p_*})}$$

from Theorem 8.3 by a value that does not depend on z; therefore, this function also generates the space $BA(p_0, z, D, x)$.

Besides,

$$f(z, x) = \frac{\psi(z, x)}{\exp\left(\sum\limits_{i=1}^{\infty} x_i z^i\right)} = \sum_{j=0}^{\infty} \xi_j(x) z^{-j},$$

with $\xi_0(x) = f(\infty, x) = 1$.

Therefore, $\psi(z, x) = \exp\left(\sum\limits_{i=1}^{\infty} x_i z^i\right)\left(1 + \sum\limits_{j=1}^{\infty} \xi_i z^{-j}\right)$.

The function from Lemma 8.4 will be called a *normalized* Baker–Akhiezer function.

Lemma 8.5 *Every normalized Baker–Akhiezer function $\psi(z, x)$ is a formal exponential, i.e., for every $n > 1$ there exists an operator*

$$L_n = \partial^n + \sum_{i=2}^{n} B_n^i(x)\partial^{n-i} \quad \text{such that} \quad \partial_n \psi = L_n \psi.$$

Proof It follows from Lemma 8.4 that

$$\partial_n \psi = z^n \exp\Big(\sum_{i=1}^{\infty} x_i z^i\Big)\Big(1 + \sum_{j=1}^{\infty} \xi_j z^{-j}\Big) + \exp\Big(\sum_{i=1}^{\infty} x_i z^i\Big)\Big(\sum_{j=1}^{\infty} \partial_n \xi_j z^{-j}\Big)$$

and

$$\partial^n \psi$$

$$= z^n \exp\Big(\sum_{i=1}^{\infty} x_i z^i\Big)\Big(1 + \sum_{j=1}^{\infty} \xi_j z^{-j}\Big) + \sum_{r=1}^{n} z^{n-r} \exp\Big(\sum_{i=1}^{\infty} x_i z^i\Big)\Big(c_r \sum_{j=1}^{\infty} \partial^r \xi_j z^{-j}\Big).$$

Hence there exist functions $B_i^r(x)$ such that

$$\partial_n \psi = \partial^n \psi + \sum_{r=2}^{i} B_n^r \partial^{n-r} \psi + \exp\Big(\sum_{i=1}^{\infty} x_i z^i\Big)\Big(\sum_{j=1}^{\infty} \hat{\xi}_j z^{-j}\Big)$$

where $\hat{\xi}_j$ are analytic functions of x. The last term satisfies the axioms for a Baker–Akhiezer function and therefore, by Lemma 8.4, is proportional to the function

$$\psi(z, x) = \exp\Big(\sum_{i=1}^{\infty} x_i z^i\Big)\Big(1 + \sum_{j=1}^{\infty} \xi_i z^{-j}\Big).$$

Thus, this term vanishes.

Example 8.1 We have

$$\partial \psi = z \exp\Big(\sum_{i=1}^{\infty} x_i z^i\Big)\Big(1 + \sum_{j=1}^{\infty} \xi_j z^{-j}\Big) + \exp\Big(\sum_{i=1}^{\infty} x_i z^i\Big)\Big(\sum_{j=1}^{\infty} \partial \xi_j z^{-j}\Big)$$

$$= \exp\Big(\sum_{i=1}^{\infty} x_i z^i\Big)\Big(z + \xi_1 + \sum_{j=1}^{\infty} (\xi_{j+1} + \partial \xi_j) z^{-j}\Big),$$

$$\partial^2 \psi = \exp\Big(\sum_{i=1}^{\infty} x_i z^i\Big)\Big(z^2 + z\xi_1 + (\xi_2 + 2\partial \xi_1)$$

$$+ \sum_{j=1}^{\infty} (\xi_{j+2} + 2\partial \xi_{j+1} + \partial^2 \xi_j) z^{-j}\Big),$$

$$\partial_2 \psi = \exp\Big(\sum_{i=1}^{N} x_i z^i\Big)\Big(z^2 + z\xi_1 + \xi_2 + \sum_{j=1}^{\infty} (\xi_{j+2} + \partial_2 \xi_j) z^{-j}\Big).$$

Thus,

$$\partial_2 \psi = \partial^2 \psi - 2\partial \xi_1 \psi + \exp\left(\sum_{i=1}^{\infty} x_i z^i\right)\left(\sum_{j=1}^{\infty} \hat{\xi}_i z^{-j}\right),$$

i.e., $B_2^2 = -2\,\partial \xi_1$.

Exercise 8.3 Show that $B_3^2 = -3\partial \xi_2$, $B_3^3 = 3\xi_1\partial \xi_1 + 3\partial^2 \xi_1 - 3\partial \xi_2$.

8.6 Algebro-Geometric Solutions of the KP and n-KdV Equations

Theorem 8.4 *The function*

$$\tau(x) = \theta\left(A_{p_*}(p_0) - A_{p_*}(D) + \sum_{i=1}^{\infty} x_i U^i - K_{p_*}\right)$$

is an almost tau function for the KP hierarchy.

Proof By Riemann's bilinear relations (Theorem 6.8), the function $A_{p_*}(p)$ in the local chart z has the form $A_{p_*}(p) = A_{p_*}(p_0) - \sum_{i=1}^{\infty} \frac{1}{i} z^{-i} U^i$. Thus, by Lemma 8.5, the function

$$\psi(z, x) = \exp\left(\int_{p_*}^{p} \sum_{i=1}^{\infty} x_i \Omega^i\right) \frac{\tau(x_1 - z^{-1}, x_2 - \frac{1}{2}z^{-2}, x_3 - \frac{1}{3}z^{-3}, \ldots)\,\tau(0)}{\tau(x_1, x_2, x_3, \ldots)\,\tau(-z^{-1}, -\frac{1}{2}z^{-2}, -\frac{1}{3}z^{-3}, \ldots)}$$

is a normalized Baker–Akhiezer function and a formal exponential. Besides, according to our definitions,

$$\int_{p_*}^{p} \Omega^j = z^i + \sum_{i=1}^{\infty} a_{ij} z^{-i}. \tag{8.8}$$

Theorem 8.5 *Let D be a nonspecial divisor on a Riemann surface P. Then the function*

$$v = \tau(x) = \ln \theta\left(\sum_{i=1}^{\infty} x_i U^i - (A_{p_0}(D) + K_{p_0})\right)$$

satisfies the KP hierarchy (8.3), and the function

$$u = 2\,\partial^2\tau(x) = 2\,\partial^2 \ln\theta\Big(\sum_{i=1}^{3} x_i U^i - (A_{p_0}(D) + K_{p_0})\Big)$$

is a solution to the KP equation.

Proof By (8.7) and (8.8), the constants a_{ij} approach 0 as p_* approaches p_0. Hence the theorem follows from Theorem 8.2 and Corollary 8.1.

Remark 8.1 Theorem 8.5 can be regarded as a differential equation for a theta function. The theorem says that the KP equation is satisfied if the theta function is constructed from a Riemann surface. In the late 1970s, S. P. Novikov conjectured that every theta function satisfying the KP equation comes from some Riemann surface. In the 1980s, this conjecture was proved by Shiota (see [25]). Thus, the KP hierarchy solves the Schottky problem of describing the theta functions of Riemann surfaces.

Now consider a pair (P, f) consisting of a Riemann surface P and a meromorphic function $f\colon P \to \mathbb{C}$ with a unique pole at a point p_0. Let n be the order of this pole and z^{-1} be a local chart in a neighborhood of p_0 in which f has the form $z \mapsto z^n$. Then $\Omega^n = df$ and, consequently, $U^n = 0$. In this case, v does not depend on x_n and, therefore, is a solution to the n-KdV hierarchy.

For $n = 2$, this construction gives a solution to the classical KdV equation. Let us describe it in more detail. The existence on a surface P of a function of degree 2 means that P is a hyperelliptic Riemann surface, i.e., there exists a holomorphic involution $\alpha\colon P \to P$ such that $P/\langle\alpha\rangle$ is the Riemann sphere. The involution has $2g + 2$ fixed points, where g is the genus of P. Consider an arbitrary fixed point p_0 and a local chart $z^{-1}\colon U \to \mathbb{C}$ in a neighborhood of p_0 such that $z^{-1}(p_0) = \infty$ and $z^{-1}(\alpha p) = -z^{-1}(p)$. Consider a nonspecial divisor D. Then, by Theorem 8.5, the function

$$u = 2\,\partial^2 \ln\theta(x_1 U^1 + x_3 U^3 - (A_{p_*}(D) + K_{p_*}))$$

satisfies the Korteweg–de Vries (KdV) equation

$$\partial_3 u = \frac{1}{4}\,(6\,u\,\partial u + \partial^3 u).$$

Algebro-geometric solutions of the KP and KdV equations have the form

$$u(x_1, x_2, x_3) = P(L(x_1, x_2, x_3))$$

where L is a linear map to $\mathbb{R}^n$ and $P(y_1, \dots, y_n)$ is a periodic function. Such functions are called *quasi-periodic*. They play a very important role in the description of wave processes.

Chapter 9
Formula for a Conformal Map from an Arbitrary Domain onto Disk

9.1 The Space of Simply Connected Domains

Conformal maps are essential in a broad range of applied problems (aeromechanics and hydromechanics, oil production, etc.). So, we would like to refine Riemann's theorem on the existence of a conformal map from a simply connected domain D to the unit disk Λ by providing an explicit construction of such a map.

Consider the space $\mathcal{H}$ of all simply connected domains with analytic boundary on the Riemann sphere that contain ∞ and whose closure does not contain the origin.

For coordinates in this infinite-dimensional space, we will take *Richardson's harmonic moments*, introduced in the late twentieth century as a tool for solving the inverse problem in potential theory:

$$
t_0 = \frac{1}{\pi} \iint\limits_{\bar{\mathbb{C}} \backslash Q} d^2 z, \quad t_k = -\frac{1}{\pi k} \iint\limits_{Q} z^{-k} d^2 z \quad (k = 1, 2, \ldots), \quad d^2 z = dx \, dy.
$$

One can prove [28, Chap. 2] that these functions are local coordinates on the space $\mathcal{H}$. We will consider functions on $\mathcal{H}$ that are not holomorphic in $\{t_k\}$. For this reason, it will be convenient to assume (as we did earlier in similar cases) that such a function can be expanded in a series in the variables $\{t_k\}$ and $\{\bar{t}_k\}$. By $Q^t \in \mathcal{H}$ we will denote the domain corresponding to the coordinates $t = \{t_0, t_1, t_2, \ldots\}$.

Denote by $\mathcal{H}_z \subset \mathcal{H}$ the set of domains containing a point $z \in \mathbb{C}$. Consider a domain $Q \in \mathcal{H}_z$, whose points will be denoted by ξ. Denote by $G_Q(z, \xi)$ the Green's function of this domain. With Q we associate the domain Q_ε obtained from Q by shifting the boundary by $-\varepsilon \pi \frac{\partial}{\partial n} G_Q(z, \xi)$ in the direction of the outer normal to the boundary of Q.

© Springer Nature Switzerland AG 2019
S. M. Natanzon, *Complex Analysis, Riemann Surfaces and Integrable Systems*,
Moscow Lectures 3, https://doi.org/10.1007/978-3-030-34640-9_9

Vector fields on $\mathcal{H}_z$ are linear functionals on the set of functions defined on $\mathcal{H}_z$. Denote by δ_z the vector field on $\mathcal{H}_z$ sending a function $X \colon \mathcal{H}_z \to \mathbb{C}$ to the function that takes on Q the value $\lim\limits_{\varepsilon \to 0} \frac{1}{\varepsilon^2} \left(X(Q_\varepsilon) - X(Q) \right)$.

Exercise 9.1 Show that $\delta_z(t_0) = 1$.

Functions on $\mathcal{H}$ can be regarded as series of variables $t_0, t_1, \bar{t}_1, t_2, \bar{t}_2, \ldots$. This allows us to consider partial derivatives with respect to these variables. Consider the following family, parametrized by z and $\bar{z}$, of differential operators in the coordinates $t = \{t_0, t_1, t_2, \ldots\}$ on $\mathcal{H}$:

$$D(z) = \sum_{k \geq 1} \frac{z^{-1}}{k} \frac{\partial}{\partial t_i}, \quad \bar{D}(\bar{z}) = \sum_{k \geq 1} \frac{\bar{z}^{-1}}{k} \frac{\partial}{\partial \bar{t}_i}, \quad \nabla(z) = \frac{\partial}{\partial t_0} + D(z) + \bar{D}(\bar{z}).$$

Lemma 9.1 *The linear functionals δ_z and $\nabla(z)$ coincide on the set $\{X\}$ of functions on $\mathcal{H}_z$. Besides, $\delta_z X = -\pi f(z)$ for a function of the form $X(Q) = \iint\limits_Q f \, d^2\xi$ generated by a harmonic function f on Q.*

Proof Let $X(Q) = \iint\limits_Q f \, d^2\xi$. Then

$$\delta_z X = \lim_{\varepsilon \to 0} \frac{1}{\varepsilon^2} \iint\limits_{Q_\varepsilon \backslash Q} f(\xi)(-\varepsilon\pi) \frac{\partial}{\partial n} G_Q(z, \xi) \, d^2\xi$$

$$= -\pi \oint\limits_{\xi \in \partial Q} f(\xi) \frac{\partial}{\partial n} G_Q(z, \xi) \, ds = -\pi f(z).$$

The coordinates t_i and $\bar{t}_i$ for $i > 0$ are also functions on $\mathcal{H}_z$ of the form under consideration. Hence

$$\delta_z(t_k) = \frac{z^{-k}}{k}, \qquad \delta_z(\bar{t}_k) = \frac{\bar{z}^{-k}}{k}.$$

It follows that

$$\delta_z(X) = \frac{\partial X}{\partial t_0} \delta_z t_0 + \sum \frac{\partial X}{\partial t_k} \delta_z t_k + \sum \frac{\partial X}{\partial \bar{t}_k} \delta_z \bar{t}_k$$

$$= \left(\frac{\partial}{\partial t_0} + \sum_{k \geq 1} \frac{z^{-k}}{k} \frac{\partial}{\partial t_k} + \sum_{k \geq 1} \frac{\bar{z}^{-k}}{k} \frac{\partial}{\partial \bar{t}_k} \right) X.$$

Exercise 9.2 Let X be a function of the form $X(Q) = \iint\limits_{\bar{\mathbb{C}} \setminus Q} f \, d^2\xi$ where f is an arbitrary domain-independent integrable function regular on the boundary. Then $\nabla(z)X = \pi f^h(z)$ where $f^h(z)$ is the harmonic function in Q that coincides with f on ∂Q.

Set

$$F(t) = -\frac{1}{\pi^2} \iint\limits_{Q_o^t} \iint\limits_{Q_o^t} \ln|z^{-1} - \xi^{-1}| \, d^2z \, d^2\xi \quad \text{where } Q_o^t = \bar{\mathbb{C}} \setminus Q^t. \tag{9.1}$$

Theorem 9.1 *The Green's function $G_Q(z, \xi)$ in the domain $Q = Q^t$ can be expressed in terms of the function $F = F(t)$ by the formula*

$$G_Q(z, \xi) = \frac{1}{2\pi} \ln|z^{-1} - \xi^{-1}| + \frac{1}{4\pi} \nabla(z)\nabla(\xi)F.$$

Proof Fix a point $z \in Q$. Using Exercise 9.2, we obtain

$$\nabla(z)F = -\frac{2}{\pi} \iint\limits_{Q_o} \ln|z^{-1} - \xi^{-1}| \, d^2\xi.$$

Applying Exercise 9.2 once again, we see that $\nabla(\xi)\nabla(z)F$ is a harmonic function on Q coinciding with the function $-2 \ln|z^{-1} - \xi^{-1}|$ on ∂Q. Thus, $G_Q(z, \xi)$ satisfies all conditions for a Green's function of the domain Q.

9.2 Conformal Maps and Integrable Systems

Denote by $w(z, t)$ the biholomorphic map from the domain Q^t to the exterior $\{z \in \bar{\mathbb{C}} \mid |z| > 1\}$ of the unit disk normalized so as to satisfy the conditions $w^t(\infty) = \infty$ and $\Im \partial_z w^t(\infty) = 0$, $\Re \partial_z w^t(\infty) > 0$.

The map has the form $w(z, t) = p(t)z + \sum\limits_{j=0}^{\infty} p_j(t)z^{-j}$ where, due to the normalization, $p(t) \in \mathbb{R}$ and $p'(0) > 0$. In this subsection, we will find the functions $p(t), p_0(t), p_1(t), \ldots$.

Exercise 9.3 Show that $w(z, t) = \sqrt{t_0}\, z$ for $t = (t_0, 0, 0, \ldots)$.

We will need the integrable system called the *two-dimensional dispersionless Toda lattice*. This is an infinite system of differential equations with special properties constructed in the late 1990s for the needs of mathematical physics. For our purposes, it will be convenient to describe it as a system of relations between the partial derivatives of a function $F(t)$. A special role is played by the derivative $\partial_0 = \frac{\partial}{\partial t_0}$. This system has the form

$$(z - \xi)\, e^{D(z)D(\xi)F} = ze^{-\partial_0 D(z)F} - \xi e^{-\partial_0 D(\xi)F},$$

$$(\bar{z} - \bar{\xi})\, e^{\bar{D}(\bar{z})\bar{D}(\bar{\xi})F} = \bar{z}e^{-\partial_0 \bar{D}(\bar{z})F} - \bar{\xi}e^{-\partial_0 \bar{D}(\bar{\xi})F},$$

$$1 - e^{-D(z)\bar{D}(\bar{\xi})F} = \frac{1}{z\bar{\xi}}\, e^{\partial_0 (\partial_0 + D(z) + \bar{D}(\bar{\xi}))F}.$$

Its solutions are functions $F(t)$ in infinitely many variables $t = (t_0, t_1, t_2, \ldots)$ satisfying these differential equations for any pair of complex numbers (z, ξ).

Theorem 9.2 *The function $F(t)$ defined by (9.1) is a solution to the two-dimensional dispersionless Toda lattice. Besides, the biholomorphic map $w(z, t)$ defined above has the form*

$$w(z, t) = z \exp\left(\left(-\frac{1}{2}\partial_0^2 - \partial_0 D(z)\right)F(t)\right).$$

Proof Using Theorems 4.3 and 9.1, we obtain

$$\ln\left|\frac{w(z) - w(\xi)}{1 - w(z)\,\bar{w}(\xi)}\right| = 2\pi G_\varrho(z, \xi) = \ln\left|\frac{1}{z} - \frac{1}{\xi}\right| + \frac{1}{2}\nabla(z)\nabla(\xi)F.$$

Multiplying by 2 yields

$$h = \ln\left|\frac{w(z) - w(\xi)}{1 - w(z)\,\bar{w}(\xi)}\right|^2 - \ln\left|\frac{1}{z} - \frac{1}{\xi}\right|^2 - \nabla(z)\nabla(\xi)F = 0.$$

Taking the limit as $\xi \to \infty$, we see that $-\ln|w(z)|^2 + \ln|z|^2 = \partial_0\nabla(z)F$, whence

$$\ln(p) = \lim_{z \to \infty} \ln\left|\frac{w(z)}{z}\right| = -\frac{1}{2}\partial_0^2 F.$$

Decomposing h into a sum of holomorphic, antiholomorphic, and constant parts in z, we see that the functions

$$h_1 = \ln\left(\frac{w(z) - w(\xi)}{1 - w(z)\,\bar{w}(\xi)}\right) - \ln\left(\frac{1}{z} - \frac{1}{\xi}\right) - D(z)\nabla(\xi)F$$

and

$$h_2 = \ln\left(\frac{\bar{w}(\bar{z}) - \bar{w}(\bar{\xi})}{1 - \bar{w}(\bar{z})\,w(\xi)}\right) - \ln\left(\frac{1}{\bar{z}} - \frac{1}{\bar{\xi}}\right) - \bar{D}(\bar{z})\nabla(\xi)F$$

do not depend on z. Taking the limit as $z \to \infty$, we obtain

$$h_1 = \ln\left(-\frac{1}{\bar{w}(\xi)}\right) - \ln\left(-\frac{1}{\xi}\right), \quad h_2 = \ln\left(-\frac{1}{w(\xi)}\right) - \ln\left(-\frac{1}{\bar{\xi}}\right).$$

Equating two expressions for h_1 yields

$$D(z)\nabla(\xi)F$$

$$= \left(\ln\left(\frac{w(z)-w(\xi)}{1-w(z)\bar{w}(\xi)}\right) - \ln\left(\frac{1}{z}-\frac{1}{\xi}\right)\right) - \left(\ln\left(-\frac{1}{\bar{w}(\xi)}\right) - \ln\left(-\frac{1}{\xi}\right)\right)$$

$$= \ln\left(\frac{w(z)-w(\xi)}{1-w(z)\bar{w}(\xi)}\frac{-z\bar{w}(\xi)}{z-\xi}\right) = \ln\left(\frac{w(z)-w(\xi)}{z-\xi}\right) + \ln\left(\frac{\frac{z}{w(z)}}{\frac{1}{-w(z)\bar{w}(\xi)}+1}\right).$$

Taking the limit as $\xi \to \infty$, we see that

$$D(z)\partial_0 F = \ln(p) + \ln\left(\frac{z}{w(z)}\right).$$

Comparing this formula with the equation $\ln(p) = -\frac{1}{2}\partial_0^2 F$ yields

$$\ln\left(\frac{w(z)}{z}\right) = -\frac{1}{2}\partial_0^2 F - D(z)\,\partial_0 F,$$

which is equivalent to

$$w(z,t) = z\exp\left(\left(-\frac{1}{2}\partial_0^2 - \partial_0 D(z)\right)F(t)\right).$$

Taking the part of the equation

$$D(z)\nabla(\xi)F = \ln\left(\frac{w(z)-w(\xi)}{z-\xi}\right) + \ln\left(\frac{\frac{z}{w(z)}}{\frac{1}{-w(z)\,\bar{w}(\xi)}+1}\right)$$

that is holomorphic with respect to ξ yields

$$-D(z)\,\partial_0 F = D(z)D(\xi)F - \ln\left(\frac{w(z)-w(\xi)}{z-\xi}\right) - \ln\left(\frac{z}{w(z)}\right)$$

$$= D(z)D(\xi)F + \ln(z-\xi) + \ln\left(\frac{w(z)}{w(z)-w(\xi)}\right) - \ln z,$$

that is,

$$ze^{-\partial_0 D(z)F} = (z - \xi)\, e^{D(z)D(\xi)F}\, \frac{w(z)}{w(z) - w(\xi)}.$$

Interchanging z and ξ shows that

$$\xi e^{-\partial_0 D(\xi)F} = (\xi - z)\, e^{D(z)D(\xi)F}\, \frac{w(\xi)}{w(\xi) - w(z)} = (z - \xi)\, e^{D(z)D(\xi)F}\, \frac{w(\xi)}{w(z) - w(\xi)}.$$

Thus,

$$ze^{-\partial_0 D(z)F} - \xi e^{-\partial_0 D(\xi)F} = (z - \xi)\, e^{D(z)D(\xi)F}.$$

Replacing (z, ξ) by $(\bar{z}, \bar{\xi})$, we obtain

$$\bar{z}e^{-\partial_0 \bar{D}(\bar{z})F} - \bar{\xi}e^{-\partial_0 \bar{D}(\bar{\xi})F} = (\bar{z} - \bar{\xi})\, e^{\bar{D}(\bar{z})\bar{D}(\bar{\xi})F}.$$

Taking the part of the equation

$$D(z)\nabla(\xi)F = \ln\left(\frac{w(z) - w(\xi)}{z - \xi}\right) + \ln\left(\frac{\frac{z}{w(z)}}{\frac{1}{-w(z)\bar{w}(\xi)} + 1}\right)$$

that depends on $\bar{\xi}$ yields

$$D(z)\bar{D}(\bar{\xi})F = -\ln\left(1 - \frac{1}{w(z)\bar{w}(\xi)}\right),$$

whence

$$e^{-D(z)\bar{D}(\bar{\xi})F} = 1 - \frac{1}{w(z)\,\bar{w}(\xi)}.$$

Substituting $w(z, t) = z\exp((-\tfrac{1}{2}\partial_0^2 - \partial_0 D(z))F(t))$, we obtain

$$1 - e^{-D(z)\bar{D}(\bar{\xi})F} = \frac{1}{z\bar{\xi}}\, e^{\partial_0(\partial_0 + D(z) + \bar{D}(\bar{\xi}))F}.$$

Exercise 9.4 Show that

$$\nabla(z)F = v_0 + 2\Re\frac{v_k}{k}z^{-1}$$

where

$$v_0 = \partial_0 F = \frac{2}{\pi} \iint\limits_{Q_\circ^t} \ln |z| \, d^2 z, \quad v_k = \frac{\partial}{\partial t_k} F = \frac{1}{\pi} \iint\limits_{Q_\circ^t} z^k \, d^2 z.$$

9.3 Formal Solutions to the Dispersionless 2D Toda Hierarhy

The second assertion of Theorem 9.2 reduces the problem of constructing a conformal map $w(z, t) \colon Q^t \to \bar{\mathbb{C}} \setminus \varLambda$ to that of explicitly calculating the function F as a series in the variables $t_0, t_1, \bar{t}_1, t_2, \bar{t}_2, \ldots$. The function $F(t)$ is also of considerable independent interest. It arises in modern models of mathematical physics (matrix models, topological gravity, etc.), where one also needs the Taylor series expansion of F. Unfortunately, the integral formula for F does not allow one to find this expansion.

The situation is rescued by the first assertion of Theorem 9.2, which says that F is a solution to the two-dimensional dispersionless Toda lattice. To look for solutions to this system, one must first exclude from it the numbers $z, \bar{z}, \xi, \bar{\xi}$. For this, one must expand the functions involved in the equations into Laurent series in $z, \bar{z}, \xi, \bar{\xi}$. The coefficients of these series are polynomials in the partial derivatives $\partial_i = \frac{\partial}{\partial t_i}$ and $\bar{\partial}_i = \frac{\partial}{\partial \bar{t}_i}$ of F. Equating the coefficients of the same monomials in $z, \xi, \bar{z}, \bar{\xi}$ in both sides of the equations yields a countable system of partial differential equations on F called the two-dimensional dispersionless Toda lattice.

Denote by $g\big|_{t_0}$ the restriction of a function $g(t_0, t_1, \bar{t}_1, t_2, \bar{t}_2, \ldots)$ to the line

$$t_1 = t_2 = \cdots = 0.$$

A solution F to the two-dimensional dispersionless Toda lattice will be called *symmetric* if $\partial_k\big|_{t_0} = \bar{\partial}_k\big|_{t_0} = 0$ for all $k > 0$. All formal (i.e., written as Taylor series which are not necessarily convergent) symmetric solutions to the two-dimensional dispersionless Toda lattice are found in [20, 21].

We begin by introducing notation convenient for our purposes. We will use *Young diagrams* $\varDelta = [\mu_1, \mu_2, \ldots, \mu_\ell]$, i.e., sequences of positive integers

$$\mu_1 \geq \mu_2 \geq \ldots \geq \mu_\ell > 0;$$

the number of terms in $\varDelta$ will be denoted by $\ell = \ell(\varDelta)$, and the sum of these terms, by $|\varDelta| = \sum_{i=1}^{\ell} \mu_i$. Set $\mathbf{t} = (t_1, t_2, \ldots)$, $t_\varDelta = t_{\mu_1} t_{\mu_2} \ldots t_{\mu_\ell}$. We introduce a similar notation also for the second group of variables: $\bar{\varDelta} = [\bar{\mu}_1, \bar{\mu}_2, \ldots, \bar{\mu}_{\bar{\ell}}]$, $\bar{\mathbf{t}} = (\bar{t}_1, \bar{t}_2, \ldots)$, $\bar{t}_{\bar{\varDelta}} = \bar{t}_{\bar{\mu}_1} \bar{t}_{\bar{\mu}_2} \ldots \bar{t}_{\bar{\mu}_{\bar{\ell}}}$.

Theorem 9.3

1° *Every formal symmetric solution to the two-dimensional dispersionless Toda lattice is determined by a single arbitrary formal function $\Phi(t_0)$ and is equal to*

$$F(t_0, \mathbf{t}, \bar{\mathbf{t}}) = \Phi(t_0) + \sum_{i>0} i f^i(t_0) t_i \bar{t}_i$$

$$+ \sum_{|\Delta|=|\bar{\Delta}|} \sum_{\substack{s_1+\ldots+s_m=|\Delta|, \\ r_1+\ldots+r_m=\ell(\Delta)+\ell(\bar{\Delta})-2>0}} N_{(\Delta|\bar{\Delta})} \begin{pmatrix} s_1 \ldots s_m \\ r_1 \ldots r_m \end{pmatrix} \partial_0^{r_1} f^{s_1}(t_0) \ldots \partial_0^{r_m} f^{s_m}(t_0) \, t_\Delta \bar{t}_{\bar{\Delta}}$$

where $f(t_0) = \exp(\partial_0^2 \Phi(t_0))$, $N_{(\Delta|\bar{\Delta})} \begin{pmatrix} s_1 \ldots s_m \\ r_1 \ldots r_m \end{pmatrix}$ are universal combinatorial constants, and the sum is taken over all Young diagrams and all positive integer indices.

2° *The coefficients $N_{(\Delta|\bar{\Delta})} \begin{pmatrix} s_1 \ldots s_m \\ r_1 \ldots r_m \end{pmatrix}$ can be found using the following scheme:*

1. *Denote by $P_{ij}(r_1,\ldots,r_m)$ the number of positive integers $(i_1,\ldots,i_m)$, $(j_1,\ldots,j_m)$ such that $i_1+\ldots+i_m = i$, $j_1+\ldots+j_m = j$, and $r_k = i_k+j_k$. Set*

$$T_{ij}(p_1,\ldots,p_m)$$

$$= \sum_{\substack{k>0,\, n_i>0, \\ n_1+\ldots+n_k=m}} \frac{(-1)^{m+1}}{k\, n_1!\ldots n_k!} P_{ij}\left(\sum_{i=1}^{n_1} p_i, \sum_{i=n_1+1}^{n_1+n_2} p_i, \ldots, \sum_{i=n_1+\ldots+n_{k-1}+1}^{m} p_i \right).$$

2. *Define a collection of numbers by the following recurrence relations:*

$$T_{i_1 i_2}\begin{pmatrix} s_1 \ldots s_m \\ \ell_1 \ldots \ell_m \end{pmatrix} = \begin{cases} T_{i_1 i_2}(s_1, \ldots, s_m) & \text{if } \ell_1 = \ldots = \ell_m = 1, \\ 0 & \text{in the other cases} \end{cases}$$

and

$$T_{i_1 \ldots i_k}\begin{pmatrix} s_1 \ldots s_m \\ \ell_1 \ldots \ell_m \end{pmatrix} = \sum_{1 \le i \le j \le m} \frac{\ell!}{(\ell_i-1)!\ldots(\ell_j-1)!}$$

$$\times T_{i_1 \ldots i_{k-1}}\begin{pmatrix} s_1 \ldots s_{i-1} \; s \; s_{j+1} \ldots s_m \\ \ell_1 \ldots \ell_{i-1} \; \ell \; \ell_{j+1} \ldots \ell_m \end{pmatrix} T_{s,i_k}(s_i, s_{i+1}, \ldots, s_j)$$

where $s = s_i + s_{i+1} + \ldots + s_j - i_k > 0$, $\ell = (\ell_i-1) + \ldots + (\ell_j-1) > 0$.

3. *Consider the numbers*

$$\tilde{N}_{\begin{pmatrix} i_1 \cdots i_k \\ \bar{i}_1 \cdots \bar{i}_{\bar{k}} \end{pmatrix}} \begin{pmatrix} s_1 \cdots s_m \\ r_1 \cdots r_m \end{pmatrix} = \frac{i_1 \ldots i_k \bar{i}_1 \ldots \bar{i}_{\bar{k}}}{s_1 \ldots s_m} \sum T_{i_1 \ldots i_k} \begin{pmatrix} s_1 & \cdots & s_m \\ r_1 - n_1 + 1 & \cdots & r_m - n_m + 1 \end{pmatrix}$$

where the sum is taken over all representations of the set $\{\bar{i}_1, \ldots, \bar{i}_{\bar{k}}\}$ as a union of nonempty subsets $\{b_1^j, \ldots, b_{n_j}^j\} \subset \{\bar{i}_1, \ldots, \bar{i}_{\bar{k}}\}$ such that $b_1^j + \ldots + b_{n_j}^j = s_j$ for $j = 1, \ldots, m$.
4. *Finally, set*

$$N_{(\Delta|\bar{\Delta})} \begin{pmatrix} s_1 \cdots s_m \\ r_1 \cdots r_m \end{pmatrix} = \frac{1}{\sigma(\Delta)\,\sigma(\bar{\Delta})} \tilde{N}_{\begin{pmatrix} \mu_1 \cdots \mu_k \\ \bar{\mu}_1 \cdots \bar{\mu}_{\bar{k}} \end{pmatrix}} \begin{pmatrix} s_1 \cdots s_m \\ r_1 \cdots r_m \end{pmatrix}$$

where Δ, $\bar{\Delta}$ are Young diagrams with rows $[\mu_1, \ldots, \mu_\ell], [\bar{\mu}_1, \ldots, \bar{\mu}_{\bar{\ell}}]$, and $\sigma(\Delta)$, $\sigma(\bar{\Delta})$ are the orders of the automorphism groups of the Young diagrams Δ and $\bar{\Delta}$.

9.4   Proof of the Theorem on Symmetric Solutions

We will prove Theorem 9.3 following [20, 21]. The proof is based on a series of lemmas.

Lemma 9.2 *The relation*

$$(z - \xi)\, e^{D(z)D(\xi)F} = z e^{-\partial_0 D(z)F} - \xi e^{-\partial_0 D(\xi)F} \tag{9.2}$$

implies

$$z - \sum_{j=1}^{\infty} \frac{1}{j} z^{-j} \partial_1 \partial_j F = z e^{-\partial_0 D(z)} F,$$

$$\partial_1 \partial_j F = \sum_{m=1}^{\infty} \frac{(-1)^{m+1}}{m!} \sum_{\substack{k_1 + \cdots + k_m = j+1 \\ k_i > 0}} \frac{j}{k_1 \ldots k_m} \partial_0 \partial_{k_1} F \ldots \partial_0 \partial_{k_m} F.$$

Proof Expanding the left-hand side of (9.2) into a Taylor series, we obtain

$$(z - \xi)e^{D(z)D(\xi)F} = (z - \xi)\left(1 + (D(z)D(\xi)F) + \frac{1}{2}(D(z)D(\xi)F)^2 + \ldots\right)$$

$$= (z - \xi)\left(1 + z^{-1}\xi^{-1}\partial_1^2 F + z^{-1}\sum_{j=2}^{\infty} \frac{1}{j}\xi^{-j}\partial_1\partial_j F\right.$$

$$+ \xi^{-1} \sum_{j=2}^{\infty} \frac{1}{j} z^{-j} \partial_1 \partial_j F + z^{-2} \xi^{-2} f \Bigg)$$

$$= (z - \xi) + \xi^{-1} \partial_1^2 F - z^{-1} \partial_1^2 F + \sum_{j=2}^{\infty} \frac{1}{j} \xi^{-j} \partial_1 \partial_j F$$

$$- \sum_{j=2}^{\infty} \frac{1}{j} z^{-j} \partial_1 \partial_j F + z^{-1} \xi^{-1} f.$$

On the other hand, according to (9.2), the function $(z - \xi) e^{D(z)D(\xi)F}$ is the sum of two functions, one depending only on z and the other depending only on ξ. Thus, $f = 0$ and $z e^{-\partial_0 D(z)F} = z - \sum_{j=1}^{\infty} \frac{1}{j} z^{-j} \partial_1 \partial_j F$. Therefore,

$$\sum_{j=1}^{\infty} \frac{1}{j} z^{-(j+1)} \partial_1 \partial_j F = 1 - e^{-\partial_0 D(z)F} = 1 - \Big(1 + \sum_{m=1}^{\infty} \frac{(-\partial_0 D(z)F)^m}{m!} \Big)$$

$$= - \sum_{m=1}^{\infty} \frac{(-1)^m}{m!} \Big(\sum_{k=1}^{\infty} \frac{z^{-k}}{k} \partial_0 \partial_k F \Big)^m$$

$$= - \sum_{m=1}^{\infty} \frac{(-1)^m}{m!} \Big(\sum_{n=1}^{\infty} z^{-n} \sum_{k_1 + \ldots + k_m = n} \frac{1}{k_1 \ldots k_m} \partial_0 \partial_{k_1} F \ldots \partial_0 \partial_{k_m} F \Big)$$

$$= - \sum_{n=1}^{\infty} z^{-n} \Big(\sum_{m=1}^{\infty} \frac{(-1)^m}{m!} \sum_{k_1 + \ldots + k_m = n} \frac{1}{k_1 \ldots k_m} \partial_0 \partial_{k_1} F \ldots \partial_0 \partial_{k_m} F \Big).$$

Thus,

$$\frac{1}{j} \partial_1 \partial_j F = - \sum_{m=1}^{\infty} \frac{(-1)^m}{m!} \sum_{k_1 + \ldots + k_m = j+1} \frac{1}{k_1 \ldots k_m} \partial_0 \partial_{k_1} F \ldots \partial_0 \partial_{k_m} F.$$

Lemma 9.3 *Relation (9.2) implies*

$$\partial_i \partial_j F$$

$$= \sum_{m=1}^{\infty} \sum_{\substack{s_1 + \cdots + s_m = i+j \\ s_i > 0}} \frac{(-1)^{m+1} ij}{m(s_1 - 1) \ldots (s_m - 1)} P_{ij}(s_1 - 1, \ldots, s_m - 1) \partial_1 \partial_{s_1 - 1} F \ldots \partial_1 \partial_{s_m - 1} F.$$

Proof By Lemma 9.2 and relation (9.2), we have

$$(z - \xi)e^{D(z)D(\xi)F} = z - \sum_{j=1}^{\infty} \frac{1}{j} z^{-j} \partial_1 \partial_j F - \left(\xi - \sum_{j=1}^{\infty} \frac{1}{j} \xi^{-j} \partial_1 \partial_j F \right)$$

$$= (z - \xi) - \sum_{j=1}^{\infty} \frac{1}{j} (z^{-j} - \xi^{-j}) \partial_1 \partial_j F.$$

Thus,

$$e^{D(z)D(\xi)} = 1 + z^{-1}\xi^{-1} \sum_{j=1}^{\infty} \frac{1}{j} \frac{(z^{-j} - \xi^{-j})}{(z^{-1} - \xi^{-1})} \partial_1 \partial_j F$$

$$= 1 + z^{-1}\xi^{-1} \sum_{j=1}^{\infty} \frac{1}{j} \left(\sum_{\substack{s+t=j-1 \\ s,t \geq 0}} z^{-s}\xi^{-t} \right) \partial_1 \partial_j F$$

$$= 1 + \sum_{j=1}^{\infty} \frac{1}{j} \left(\sum_{\substack{s+t=j+1 \\ s,t \geq 1}} z^{-s}\xi^{-t} \right) \partial_1 \partial_j F.$$

Therefore,

$$D(z)D(\xi)F = \sum_{m=1}^{\infty} \frac{(-1)^{m+1}}{m} \left(\sum_{n=1}^{\infty} \left(\sum_{\substack{s+t=n+1 \\ s,t \geq 1}} z^{-s}\xi^{-t} \right) \frac{1}{n} \partial_1 \partial_n F \right)^m$$

$$= \sum_{j=1}^{\infty} \frac{(-1)^{m+1}}{m} \sum_{i,j \geq 1} z^{-i}\xi^{-j}$$

$$\times \left(\sum_{\substack{i_1+\ldots+i_m=i \\ j_1+\ldots+j_m=j;\, i_k,j_k \geq 1}} \frac{1}{i_1 + j_1 - 1} \partial_1 \partial_{i_1+j_1-1} F \ldots \frac{1}{i_m + j_m - 1} \partial_1 \partial_{i_m+j_m-1} F \right),$$

whence

$$\partial_i \partial_j F = \sum_{m=1}^{\infty} \sum_{s_1+\ldots+s_m=i+j} \frac{(-1)^{m+1}}{m} \frac{ij}{(s_1 - 1) \ldots (s_m - 1)}$$

$$\times P_{ij}(s_1 - 1, \ldots, s_m - 1) \partial_1 \partial_{s_1-1} F \ldots \partial_1 \partial_{s_m-1} F.$$

Remark 9.1 The family of differential equations

$$\partial_i \partial_j F$$

$$= \sum_{m=1}^{\infty} \sum_{\substack{s_1+\cdots+s_m=i+j \\ s_i>0}} \frac{(-1)^{m+1}ij}{m(s_1-1)\dots(s_m-1)} P_{ij}(s_1-1,\dots,s_m-1)\partial_1\partial_{s_1-1}F\dots\partial_1\partial_{s_m-1}F$$

is the dispersionless limit of the KP hierarchy. Another description of this limit is suggested in [15]. Comparing these two descriptions leads to nontrivial combinatorial relations between the combinatorial constants $P_{i,j}$.

Lemma 9.4 *Relation* (9.2) *implies*

$$\partial_i \partial_j F = \sum_{m=1}^{\infty} \sum_{p_1+\dots+p_m=j+i} \frac{ij}{p_1\dots p_m} T_{ij}(p_1\dots p_m)\partial_0\partial_{p_1}F\dots\partial_0\partial_{p_m}F.$$

Proof By Lemmas 9.2 and 9.3, we have

$$\partial_i \partial_j F = \sum_{m=1}^{\infty} \sum_{s_1+\dots+s_m=j+i} \frac{(-1)^{m+1}}{m} \frac{ij}{(s_1-1)\dots(s_m-1)} P_{ij}(s_1-1,\dots,s_m-1)$$

$$\times \partial_1\partial_{s_1-1}F\dots\partial_1\partial_{s_m-1}F) = \sum_{m=1}^{\infty} \sum_{s_1+\dots+s_m=j+i} \frac{(-1)^{m+1}}{m} \frac{ij}{(s_1-1)\dots(s_m-1)}$$

$$\times P_{ij}(s_1-1,\dots,s_m-1)\Big(\sum_{n_1=1}^{\infty} \sum_{p_1+\dots+p_{n_1}=s_1} \frac{(-1)^{n_1+1}}{n_1!} \frac{s_1-1}{p_1\dots p_{n_1}}\partial_0\partial_{p_1}F\dots\partial_0\partial_{p_{n_1}}F\Big)\dots$$

$$\dots\Big(\sum_{n_m=1}^{\infty} \sum_{p_1+\dots+p_{n_m}=s_m} \frac{(-1)^{n_m+1}}{n_m!} \frac{s_m-1}{p_1\dots p_{n_m}}\partial_0\partial_{p_1}F\dots\partial_0\partial_{p_{n_m}}F\Big)$$

$$= \sum_{m=1}^{\infty} \sum_{p_1+\dots+p_m=j+i} \frac{ij}{p_1\dots p_m} T_{ij}(p_1\dots p_m)\partial_0\partial_{p_1}F\dots\partial_0\partial_{p_m}F.$$

The next result follows from Lemma 9.4 by induction on k.

Lemma 9.5 *Relation* (9.2) *implies*

$$\partial_{i_1} \partial_{i_2} \dots \partial_{i_k} F$$

$$= \sum_{m=1}^{\infty} \Big(\sum_{\substack{s_1+\dots+s_m=i_1+\dots+i_k \\ \ell_1+\dots+\ell_m=m+k-2}} \frac{i_1\dots i_k}{s_1\dots s_m} T_{i_1\dots i_k}\begin{pmatrix} s_1\dots s_m \\ \ell_1\dots \ell_m \end{pmatrix}\partial_0^{\ell_1}\partial_{s_1}F\dots\partial_0^{\ell_m}\partial_{s_m}F\Big).$$

Now consider an arbitrary symmetric formal solution F to the two-dimensional dispersionless Toda lattice and set $f = \exp(F|_{t_0}'')$.

Lemma 9.6 *Symmetric formal solutions F satisfy the relation*

$$
\partial_i \bar{\partial}_j F|_{t_0} =
\begin{cases}
0 & \text{if } i \neq j, \\
i f^i & \text{if } i = j.
\end{cases}
$$

Proof For $k > 0$, the relation $\partial_0 \partial_k F\big|_{t_0} = \partial_0 \bar{\partial}_k F\big|_{t_0} = 0$ implies

$$
\exp(\partial_0(\partial_0 + D(z) + \bar{D}(\bar{\xi}))F)\Big|_{t_0} = \exp(\partial_0^2 F|_{t_0}) = \exp(F|_{t_0}'') = f.
$$

Hence, the relation $1 - e^{-D(z)\bar{D}(\bar{\xi})F} = z^{-1}\bar{\xi}^{-1} e^{\partial_0(\partial_0 + D(z) + \bar{D}(\bar{\xi}))F}$ implies

$$
-D(z)\bar{D}(\bar{\xi})F\Big|_{t_0} = \log(1 - z^{-1}\bar{\xi}^{-1} f) = -\sum_{k=1}^{\infty} \frac{1}{k} z^{-k} \bar{\xi}^{-k} f^k.
$$

Therefore, $\partial_i \bar{\partial}_j F\big|_{t_0} = 0$ for $i \neq j$ and $\partial_i \bar{\partial}_i F\big|_{t_0} = i f^i$.

Lemma 9.7 *Symmetric formal solutions F satisfy the relation*

$$
\partial_i \bar{\partial}_{i_1} \ldots \bar{\partial}_{i_k} F\Big|_{t_0} = \bar{\partial}_i \partial_{i_1} \ldots \partial_{i_k} F\Big|_{t_0} =
\begin{cases}
0 & \text{if } i_1 + \ldots + i_k \neq i, \\
i_1 \ldots i_k \partial_0^{k-1}(f^i) & \text{if } i_1 + \ldots + i_k = i.
\end{cases}
$$

Proof The differentials ∂ and $\bar{\partial}$ enter the system of equations symmetrically. This implies the first relation. Besides, it follows from Lemma 9.7 that

$$
\bar{\partial}_i \partial_{i_1} \partial_{i_2} \ldots \partial_{i_k} F = \frac{i_1 \ldots i_k}{i_1 + \ldots + i_k} \partial_0^{k-1} \bar{\partial}_i \partial_{i_1 + \ldots + i_k} F
$$

$$
+ \bar{\partial}_i \sum_{\substack{m=2 \\ s_1 + \ldots + s_m = i_1 + \ldots + i_k \\ \ell_1 + \ldots + \ell_m = m + k - 2}}^{\infty} \sum_{s_1 \ldots s_m} \frac{i_1 \ldots i_k}{s_1 \ldots s_m} T_{i_1 \ldots i_k} \binom{s_1 \ldots s_m}{\ell_1 \ldots \ell_m} \partial_0^{\ell_1} \partial_{s_1} F \ldots \partial_0^{\ell_m} \partial_{s_m} F.
$$

This relation, together with Lemma 9.6, implies the second relation of Lemma 9.7.

Lemma 9.8 *Symmetric formal solutions F for $k, \bar{k} > 1$ satisfy the relation*

$$
\partial_{i_1} \ldots \partial_{i_k} \bar{\partial}_{\bar{i}_1} \ldots \bar{\partial}_{\bar{i}_{\bar{k}}} F\Big|_{t_0}
$$

$$
= \sum_{\substack{m=1 \\ s_1 + \ldots + s_m = i_1 + \ldots + i_k = \bar{i}_1 + \ldots + \bar{i}_{\bar{k}} \\ r_1 + \ldots + r_m = k + \bar{k} - 2}}^{\infty} \sum \tilde{N}_{\binom{i_1 \ldots i_k}{\bar{i}_1 \ldots \bar{i}_{\bar{k}}}} \binom{s_1 \ldots s_m}{r_1 \ldots r_m} \partial_0^{r_1} f^{s_1} \ldots \partial_0^{r_m} f^{s_m}.
$$

Proof By Lemma 9.5, we have

$$\partial_{i_1} \ldots \partial_{i_k} \bar{\partial}_{\bar{i}_1} \ldots \bar{\partial}_{\bar{i}_{\bar{k}}} F$$

$$= \bar{\partial}_{\bar{i}_{\bar{i}}} \ldots \bar{\partial}_{\bar{i}_{\bar{k}}} \left(\sum_{m=1}^{\infty} \sum_{\substack{s_1+\ldots+s_m=i_1+\ldots+i_k \\ l_1+\ldots+l_m=m+k-2}} \frac{i_1 \ldots i_k}{s_1 \ldots s_m} \cdot T_{i_1 \ldots i_k} \begin{pmatrix} s_1 \ldots s_m \\ l_1 \ldots l_m \end{pmatrix} \partial_0^{l_1} \partial_{s_1} F \ldots \partial_0^{l_m} \partial_{s_m} F \right)$$

$$= \sum_{m=1}^{\infty} \sum_{\substack{s_1+\ldots+s_m=i_1+\ldots+i_k \\ l_1+\ldots+l_m=m+k-2}} \sum \left(\frac{i_1 \ldots i_k}{s_1 \ldots s_m} T_{i_1 \ldots i_k} \begin{pmatrix} s_1 \ldots s_m \\ l_1 \ldots l_m \end{pmatrix} \right.$$

$$\left. \times \partial_0^{l_1} \partial_{s_1} \bar{\partial}_{\bar{j}_1^1} \ldots \bar{\partial}_{\bar{j}_{n_1}^1} F \ldots \partial_0^{l_m} \partial_{s_m} \bar{\partial}_{\bar{j}_1^m} \ldots \bar{\partial}_{\bar{j}_{n_m}^m} F \right)$$

where the last sum is taken over all representations of the set $\{\bar{i}_1, \ldots, \bar{i}_{\bar{k}}\}$ as a union of nonempty subsets $\{b_1^j, \ldots, b_{n_j}^j\} \subset \{\bar{i}_1, \ldots, \bar{i}_{\bar{k}}\}$ such that $b_1^j + \ldots + b_{n_j}^j = s_j$ for $j = 1, \ldots, m$.

Therefore, by Lemma 9.7,

$$\partial_{i_1} \ldots \partial_{i_k} \bar{\partial}_{\bar{i}_1} \ldots \bar{\partial}_{\bar{i}_{\bar{k}}} F$$

$$= \sum_{m=1}^{\infty} \sum_{\substack{s_1+\ldots+s_m=i_1+\ldots+i_k \\ l_1+\ldots+l_m=m+k-2}} \frac{i_1 \ldots i_k \, \bar{i}_1 \ldots \bar{i}_{\bar{k}}}{s_1 \ldots s_m} \sum \left(T_{i_1 \ldots i_k} \begin{pmatrix} s_1 \ldots s_m \\ l_1 \ldots l_m \end{pmatrix} \partial_0^{l_1+n_1-1} (f^{s_1}) \times \ldots \right.$$

$$\left. \ldots \times \partial_0^{l_m+n_m-1} (f^{s_m}) \right)$$

$$= \sum_{m=1}^{\infty} \sum_{\substack{s_1+\ldots+s_m=i_1+\ldots+i_k \\ r_1+\ldots+r_m=k+\bar{k}-2}} \frac{i_1 \ldots i_k \, \bar{i}_1 \ldots \bar{i}_{\bar{k}}}{s_1 \ldots s_m}$$

$$\times \sum \left(T_{i_1 \ldots i_k} \begin{pmatrix} s_1 \ldots s_m \\ r_1 - n_1 + 1 \ldots r_m - n_m + 1 \end{pmatrix} \partial_0^{r_1} (f^{s_1}) \ldots \partial_0^{r_m} (f^{s_m}) \right)$$

where the last sum is taken over all representations of the set $\{\bar{i}_1, \ldots, \bar{i}_{\bar{k}}\}$ as a union of nonempty subsets $\{b_1^j, \ldots, b_{n_j}^j\} \subset \{\bar{i}_1, \ldots, \bar{i}_{\bar{k}}\}$ such that $b_1^j + \ldots + b_{n_j}^j = s_j$ for $j = 1, \ldots, m$.

The assertion of Theorem 9.3 is equivalent to Lemmas 9.7 and 9.8.

9.5 Effectivization of Riemann's Theorem

One can easily show that the solution to the two-dimensional dispersionless Toda lattice that produces conformal maps is symmetric. By Exercise 9.3, the corresponding function $\Phi(t_0)$ can be found using the condition

$$\exp((-\frac{1}{2}\partial_0^2 - \partial_0 D(z))\,F(t)) = \sqrt{t_0} \quad \text{for } t = (t_0, 0, 0, \dots).$$

Therefore,

$$\Phi(t_0) = \frac{1}{2} t_0^2 \ln t_0 - \frac{3}{4} t_0^2, \quad f(t_0) = t_0,$$

and the required solution is equal to

$$F(t_0, \mathbf{t}, \bar{\mathbf{t}}) = \frac{1}{2} t_0^2 \ln t_0 - \frac{3}{4} t_0^2 + \sum_{i>0} i t_0^i t_i \bar{t}_i$$

$$+ \sum_{\substack{|\Delta|=|\bar{\Delta}|\,,\, r_i \le s_i,\, s_1+\dots+s_m=|\Delta|,\\ r_1+\dots+r_m=\ell(\Delta)+\ell(\bar{\Delta})-2>0}} \sum N_{(\Delta|\bar{\Delta})} \binom{s_1 \dots s_m}{r_1 \dots r_m} \prod_{i=1}^{m} \frac{s_i!}{(s_i - r_i)!} t_0^{|\Delta|+2-\ell(\Delta)-\ell(\bar{\Delta})} t_\Delta \bar{t}_{\bar{\Delta}}.$$

Thus, we have obtained the following result.

Theorem 9.4 ([18, 20, 21]) *The conformal map* $w_Q: Q \rightarrow \Lambda$ *is given by the formula*

$$w_Q(z) = z \exp\left(\left(-\frac{1}{2}\frac{\partial^2}{\partial t_0^2} - \frac{\partial}{\partial t_0}\sum_{k \ge 1}\frac{z^{-k}}{k}\frac{\partial}{\partial t_k}\right)F\right)(t_0(Q), \mathbf{t}(Q), \bar{\mathbf{t}}(Q))$$

where

$$F(t_0, \mathbf{t}, \bar{\mathbf{t}}) = \frac{1}{2} t_0^2 \ln t_0 - \frac{3}{4} t_0^2 + \sum_{i>0} i t_0^i t_i \bar{t}_i$$

$$+ \sum_{\substack{|\Delta|=|\bar{\Delta}|\,,\, r_i \le s_i,\, s_1+\dots+s_m=|\Delta|,\\ r_1+\dots+r_m=\ell(\Delta)+\ell(\bar{\Delta})-2>0}} \sum N_{(\Delta|\bar{\Delta})} \binom{s_1 \dots s_m}{r_1 \dots r_m} \prod_{i=1}^{m} \frac{s_i!}{(s_i - r_i)!} t_0^{|\Delta|+2-\ell(\Delta)-\ell(\bar{\Delta})} t_\Delta \bar{t}_{\bar{\Delta}}.$$

The convergence of the formal function $w_Q(z)$ from Theorem 9.4 remains an important open problem. Progress in solving this problem was achieved in [8], namely, sufficient conditions were obtained for the convergence of the formal series $F(t_0, \mathbf{t}, \bar{\mathbf{t}})$ from Theorem 9.4.

Theorem 9.5 *Let $\tilde{t} = (t, \boldsymbol{t}, \bar{\boldsymbol{t}})$ where $t_i, \bar{t}_i = 0$ for $i > n$, $0 < t_0 < 1$, and*

$$4|t_i|, |\bar{t}_i| \leq (4n^3 2^n e^n)^{-1}.$$

Then the series defining $F(t, \boldsymbol{t}, \bar{\boldsymbol{t}})$ in Theorem 9.4 converges.

References

1. Date, E., Kashiwara, M., Jimbo, M., Miwa, T.: Transformation groups for soliton equation. In: Jimbo, M., Miwa, T. (eds.). Proceedings of RIMS Symposium on Non-Linear Integrable Systems, pp. 39–119. World Science, Singapore (1983)
2. Dubrovin, B.A.: Theta functions and non-linear equations. Russ. Math. Surv. **36**(2), 11–92 (1981)
3. Dubrovin, B.A.: Riemann Surfaces and Nonlinear Equations [in Russian]. RHD, Moscow–Izhevsk (2001)
4. Dubrovin, B.A., Natanzon S.M.: Real theta-function solutions of the Kadomtsev–Petviashvili equation. Math. USSR Izv. **32**(2), 269–288 (1989)
5. Fricke, R., Klein, F.: Vorlesungen über die Theorie der automorphen Funktionen. Bd. 2. Teubner, Leipzig (1897, 1912)
6. Griffiths, P., Harris, J.: Principles of Algebraic Geometry. Wiley, New York (1978)
7. Its, A.R., Matveev, V.B.: Schrödinger operators with finite-gap spectrum and N-soliton solutions of the Korteweg–de Vries equation. Theor. Math. Phys. **23**(1), 343–355 (1975)
8. Klimov, Yu., Korzh, A., Natanzon, S.: From 2D Toda hierarchy to conformal maps for domains of the Riemann sphere. Am. Math. Soc. Transl. (2) **212**, 207–218 (2004)
9. Kostov, I., Krichever, I., Mineev-Weinstein, M., Wiegmann, P.B., Zabrodin, A.: The τ-function for analytic curves. In: Random Matrix Models and Their Applications. Mathematical Sciences Research Institute Publications, vol. 40, pp. 285–299. Cambridge University Press, Cambridge (2001)
10. Krichever, I.M.: Algebraic-geometric construction of the Zakharov–Shabat equations and their periodic solutions. Sov. Math. Dokl. **17**, 394–397 (1976)
11. Krichever, I.M.: Methods of algebraic geometry in the theory of nonlinear equations. Russian Math. Surveys **32**(6), 185–213 (1977)
12. Mineev-Weinstein, M., Wiegmann, P.B., Zabrodin, A.: Integrable structure of interface dynamics. Phys. Rev. Lett. **84**(22), 5106–5109 (2000)
13. Natanzon, S.M.: Invariant lines of Fuchsian groups. Russ. Math. Surv. **27**(4), 161–177 (1972)
14. Natanzon, S.M.: Moduli spaces of real curves. Tr. Mosk. Mat. Obs. **37**, 219–253 (1978)
15. Natanzon, S.M.: Formulas for A_n- and B_n-solutions of WDVV equations. J. Geom. Phys. **39**(4), 323–336 (2001)
16. Natanzon, S.M.: Witten solution for the Gelfand–Dikii hierarchy. Funct. Anal. Appl. **37**(1), 21–31 (2003)
17. Natanzon, S.M.: Moduli of Riemann Surfaces, Real Algebraic Curves, and Their Superanalogs. American Mathematical Society, Providence (2004)

© Springer Nature Switzerland AG 2019   135
S. M. Natanzon, *Complex Analysis, Riemann Surfaces and Integrable Systems*,
Moscow Lectures 3, https://doi.org/10.1007/978-3-030-34640-9

18. Natanzon, S.: Towards an effectivisation of the Riemann theorem. Ann. Global Anal. Geom. **28**(3), 233–255 (2005)
19. Natanzon, S.M.: A Brief Course in Mathematical Analysis. MCCME, Moscow (2008)
20. Natanzon, S.M.: Dispersionless 2D Toda hierarchy, Hurwitz numbers and Riemann theorem. J. Phys. Conf. Ser. **670**, 1–6 (2016)
21. Natanzon, S., Zabrodin, A.: Symmetric solutions to dispersionless 2D Toda hierarchy, Hurwitz numbers and conformal dynamics. Int. Math. Res. Not. **2015**(8), 2082–2110 (2015)
22. Natanzon, S.M., Zabrodin, A.V.: Formal solutions to the KP hierarchy. J. Phys. A **49**(14), 145206 (2016)
23. Novikov, S.P.: The periodic problem for the Korteweg–de Vries equation. Funct. Anal. Appl. **8**(3), 236–246 (1974)
24. Shabat, B.V.: Introduction to Complex Analysis [in Russian]. Nauka, Moscow (1969)
25. Shiota, T.: Characterization of Jacobian varieties in terms of soliton equations. Invent. Math. **83**(2), 333–382 (1986)
26. Springer, G.: Introduction to Riemann Surfaces. Addison-Wesley, Reading (1957)
27. Teichmuller, O.: Extremale quasikonforme Abbildungen und quadratische Differentiale. Abh. Preuss. Akad. Wiss. Math. Naturw. Kl. **22**, 3–197 (1940)
28. Varchenko, A.N., Etingof, P.I.: Why the Boundary of a Round Drop Becomes a Curve of Order Four. American Mathematical Society, Providence (1991)
29. Wiegmann, P.B., Zabrodin, A.: Conformal maps and integrable hierarchies. Commun. Math. Phys. **213**(3), 523–538 (2000)

Index

© Springer Nature Switzerland AG 2019
S. M. Natanzon, *Complex Analysis, Riemann Surfaces and Integrable Systems*,
Moscow Lectures 3, https://doi.org/10.1007/978-3-030-34640-9

Printed in the United States
By Bookmasters